A Pilot's Guide to Safe Flying

BY THE SAME AUTHOR

178 Seconds
Almost There
The Concise Guide to Safe Flying

A Pilot's Guide to Safe Flying

For GA, Sport and Recreational Pilots

Second Edition

Sander Vandeth

A Pilot's Guide to Safe Flying

For GA, Sport and Recreational Pilots

First Edition 2003
Second Edition 2009

Published by mCOVE Resources
PO Box 40, Mt Eliza, Victoria, Australia 3930
www.mcove.com

ISBN 978-0-9805648-0-8

Manual Design	:	Anton van Deth
Cover Photograph	:	Photo taken by Sam Bristow of his father, Bill Bristow, flying his Malibu Mirage over the Olgas (Kata Tjuta) in Central Australia. Courtesy Sam Bristow.

Table of Contents

Preface

Many pilots have succumbed to "get-thereitis". This "disease" occurs when pilots press on regardless of the obvious hazardous warning signs, usually with dire consequences. "Get-thereitis" is one example where the safety of a flight is potentially impaired by the actions or inactions of the pilot. Commonly known as pilot error. I, like most pilots, would consider myself safe, yet I have had my fair share of close calls. None resulted in an accident, but I was often left a little unnerved and wondered why I didn't see it coming or what I could have done to avoid the situation. Others have not, however, been as fortunate. With most general aviation accidents the result of human failure, the question arises: "Why do pilots make errors and what can be done differently to avoid them?"

We all know that gaining your licence is simply the beginning of a long learning journey. Each time we fly, open a book or share our experiences with fellow pilots, our knowledge base increases. In fact, the available information on flying is enormous and includes the lessons from past accidents. The surprising thing is that preventable accidents continue to occur, despite the fact that the avoidance measures have been previously identified.

It seemed to me that what might help, is a comprehensive and concise practical guide for pilots, summarizing the recognized ways of minimizing the risk of having an accident due to pilot error. This would provide a ready reference for pilots to obtain (and maintain) a core understanding of the issues involved. With improved pilot knowledge, there might hopefully be a resulting reduction in the accident rate. This guide was developed with this objective in mind. It is a manual designed for ease of use and continual reference.

I hasten to point out to the reader that the manual addresses only the key points. It should be supplemented by additional reading of the many excellent publications that are, in many cases, readily available. The Selected Bibliography and References list some of these rich sources of information.

SV

Acknowledgments

In writing this manual, I gratefully acknowledge the assistance and material provided by the regulatory authorities and other organizations.

Commonwealth Bureau of Meteorology	Cessna Aircraft Company
Civil Aviation Safety Authority of Australia	Australian Defense Force
Civil Aviation Authority of the United Kingdom	Transport Canada
Australian Transport Safety Bureau	Air BP
Civil Aviation Authority of New Zealand	Crew Training International
Aircraft Owners and Pilots Association	AOPA Air Safety Foundation
Canadian Owners and Pilots Association	Federal Aviation Administration

A number of individual people helped in various ways, and I am indebted to them for their assistance and encouragement. Their help was greatly appreciated.

Hilary Bray	Ron Chippindale	Jack and Kim Jarden
John Riley	Stuart Rushton	Sophie Wilkins

I particularly want to thank the authors and publishers who kindly allowed me to use material from their publications. They were not only gracious enough to give their permission, but were also generous in their encouragement. I also wish to acknowledge my gratitude to the various photographers who generously allowed me to use their images. Thanks to Ron Chippindale for his detailed critique of the final draft. His input was invaluable. My thanks to my son Anton and my wife Jane, for their help on the many aspects of writing, publishing and marketing this manual. Finally, my thanks to my grandson, Liam, who expertly redrew the diagram in Section 4.

Advisory

The information contained in this manual is advisory in nature, and persons using the recommendations, procedures, methods or ideas, do so assuming all responsibility and any risks thereof. It is the responsibility of the reader to determine that any recommendations, procedures, methods or ideas are not in conflict with the Airplane Flight Manual, Pilot Operating Handbook or any statutory regulations and procedures. Where such conflict may exist, then these latter requirements are applicable.

Minimizing the Risk

General aviation is a relatively safe activity provided suitable account is taken of the limits of airplanes and the fallibility of human beings. If these factors are not taken into consideration, the results can be serious (sometimes fatal) as evidenced by ongoing accidents. Most of these (approximately 80%) are the result of human related factors, many of which can be attributed to the pilot. That is, pilot error.

What can be classified as pilot error?

Here are some pilot-related factors that have contributed to accidents:

▶ Inadequate preflight preparation or planning.
▶ Improper pilot technique or procedure.
▶ Mismanagement of fuel or fuel systems.
▶ Improper in-flight decisions and planning.
▶ Lack of familiarity with the airplane.
▶ Attempted operation beyond experience and ability.
▶ Selection of an unsuitable area for takeoff or landing.
▶ Attempted visual flight into instrument meteorological conditions.
▶ Failure to see and avoid objects or obstructions.
▶ Improper loading of the airplane.
▶ Failure to extend the landing gear.
▶ Continued flight into known adverse or deteriorating weather.
▶ Self induced pressure or pressure from others.

And the list goes on!

However, accidents due to pilot error *are avoidable* if we understand how to *minimize the risk.*

Most activities we undertake in life involve some element of risk. General aviation is no exception, although the degree of risk can vary significantly depending on the circumstances and choices we make. For example, a short flight on a calm sunny day involves far less risk than a cross-country flight over high ground in extreme weather. Therefore, in order to fly safely we must be able to assess and manage risks. This involves:

- Being aware of the risks by analyzing what could go wrong. For example, by carrying out a "what if" exercise.
- Understanding the relative level of risk. That is, your ability to cope with any risks and the consequences if that is not possible.
- Taking appropriate action to ensure that the risks are minimized or even eliminated.

However, in order to minimize risk, we must in turn control errors. Pilot error can simply be defined as, an action or in-action by the pilot that results in a decrease in the margin of safety. While many errors are minor and are of little consequence, some have the impact of increasing the level of risk. For example:

- Taking a chance that the weather will improve is an error in judgment and decision making that can increase the risk of hazardous weather being encountered.
- Having an inadequate awareness of how to avoid and handle airframe icing is an error in preparation (lack of knowledge) that can increase the risk of losing control of the airplane.
- Switching to the wrong fuel tank is an error in carrying out an operational procedure that can increase the risk of fuel starvation.
- A poor landing approach is an error in proficiency that can increase the risk of a landing accident.
- Misunderstanding an ATC instruction is an error in communications that can increase the risk of a mid-air collision.
- Not detecting a drop in the manifold pressure, at an early stage, is an error in checking that can increase the risk of an engine failure.

Investigation and research has shown that most accidents are, in fact, the result of a sequence of errors - known as an error chain. As this chain grows, so does the risk of an unsafe outcome to the flight. An example of an error chain, leading to an accident, is as follows:

- Being in a hurry.
- Not obtaining a weather forecast.
- Not recognizing the hazardous weather warning signs.
- Succumbing to "get-thereitis" and pressing on despite deteriorating weather.
- Not being proficient at instrument flying and becoming disoriented in cloud.

While it is not practicable to eliminate all human error, it is possible to limit the effect of errors by limiting their occurrence and breaking the error chain. In the above example, had the hazardous weather signs been recognized and an early decision made to return, then the accident could have been avoided. Better still, had the desire to hurry been seen as a warning sign and the flight postponed, then the chain would have been cut at the very outset. As Barry Schiff noted in the

November 2000 issue of AOPA Pilot, "Being aggressively safe means that pilots should regard every out-of-the-ordinary or unexpected event, however innocuous it might appear, as the possible first link in a chain (of errors) that leads to an increasing degradation in safety. The idea is to do whatever is necessary and practical to break that link as soon as possible."

The *key* to safe flying, therefore, is to minimize human errors and break any error chain before unacceptable risks are taken. This is not as difficult as it might seem if you follow four simple strategies:

> ▸ *Have the right mental approach to flying.*
> Risk minimization starts by having the right mental approach. This involves avoiding attitudes and behaviors that are hazardous to flight such as "get-thereitis" or "it will never happen to me". Pilots must understand their limitations, and have an ability to make judgments and decisions that lead to safe outcomes, including knowing when not to fly.
>
> ▸ *Be thoroughly prepared for a flight.*
> In many ways, this is the key to safe flying. It begins with checking to see that your knowledge and proficiency matches the needs, as well as the potential needs, of the flight. This is followed by thorough, and unhurried preflight planning, where all the relevant issues, options and risks are considered and assessed. The airplane must then be inspected thoroughly with an understanding of what is being examined and why. Finally, you must be mentally prepared for the expected as well as the unexpected, so that you will be in the best position to adapt and safely cope with any situation that may arise.
>
> ▸ *Avoid "cockpit" errors.*
> "Cockpit" errors are those that arise from circumstances generated in the cockpit, such as failure to detect a fault at the earliest opportunity or becoming stressed and making errors as the result of getting "behind the airplane". Avoiding these errors requires a disciplined approach to checks, staying ahead of the airplane, avoiding distractions, avoiding communication breakdown and maintaining situational awareness.
>
> ▸ *Avoid and manage potential hazards.*
> Risks can arise from errors made in judgment, procedures, communication or proficiency, with respect to any number of possible hazards (e.g. deteriorating weather, icing, short runways, illusions, equipment failure, hypoxia, fatigue and so on). Avoiding such errors requires the application of recognized strategies and tactics for either avoiding the hazards, or, if all else fails, ensuring that any hazards are managed in the safest possible manner.

Application of these strategies provides a sound basis for error avoidance, detection and mitigation, and is discussed in detail in the subsequent sections of this manual. It permits pilots to sense when something is wrong, consider the options and take whatever corrective action is appropriate, thus breaking the error chain.

Most accidents are therefore avoidable, and pilots who *actively assess and manage flying risks* can significantly improve their level of safety.

The Right Mental Approach

2

It is an established fact that a pilot's mental and emotional makeup is a factor in many accidents. Our mental mindset frequently determines our motivation, our reaction to situations in which we find ourselves and influences our ability to make safe judgments and decisions. *The right mental approach* is therefore vital.

Avoiding Hazardous Attitudes

"One of the factors that may have contributed to this accident was that the pilot probably had a strong desire to reach the planned destination"

A pilot's behavior is generally governed by his or her attitude towards a particular situation. Attitude can be defined as a person's predisposition to act or respond in a certain way. For example, some people may have a general disregard for rules and procedures and will often ignore them if, in their opinion, it is "safe" to do so. Such an attitude, when flying, can potentially have dangerous consequences. In fact, it has been shown that certain attitudes, known as hazardous attitudes, have been linked to a significant number of general aviation accidents. These attitudes lead to poor judgments and decisions affecting both low time and experienced pilots. Furthermore, hazardous attitudes are present in all of us to some degree, and we may, therefore, need to consciously counter any adverse tendencies if we are to avoid increasing the risk of having an accident.

Hazardous Attitudes

The United States Federal Aviation Administration (FAA) has identified five hazardous attitudes to aviation, known as: anti-authority, impulsiveness, machoism, invulnerability and resignation.

Anti-authority

Anti-authority reflects resentment of supervision and authority. It is found in pilots who do not like anyone telling them what to do or simply regard rules and procedures as unnecessary.

> *"Bureaucracy is a waste of time."*
> *"Checklists are for other people."*
> *"Rules were made to be broken."*
> *"Don't tell me."*
> *"Planning is only for pilots without experience and skill."*

Impulsiveness

Impulsiveness indicates a lack of thinking before acting. Pilots with this attitude often do the first thing that comes to mind, irrespective of the alternatives.

> *"I'm sure the weather will be alright."*
> *"I must get there."*
> *"I just have to takeoff."*
> *"I must land the airplane now."*
> *"I'm in a hurry."*

Machoism

A macho attitude denotes a pilot who tries to prove they are better than anyone else or is simply showing off. They are risk takers.

> *"I'll show them."*
> *"I can make it."*
> *"I don't need any help."*
> *"I'm not going to let this beat me."*

Invulnerability

Many people, while knowing full well that accidents occur, think that it will never happen to them. Pilots who feel they are invulnerable will often take unnecessary risks and will be unprepared in an emergency.

> *"Accidents only happen to others."*
> *"It won't happen to me."*
> *"But I've done it before."*
> *"I wouldn't make such a stupid mistake."*

Resignation

Resignation reflects an attitude where pilots feel they cannot influence the outcome of things that happen to them. In many cases they consider it simply good luck or bad luck.

> *"What's the use."*
> *"It won't make any difference."*
> *"It's going to happen anyway – why fight it."*
> *"Things always work out."*
> *"I want to please."*

Hazardous attitudes, in turn, reflect themselves as unsafe behaviors. Some examples are:

- Letting others influence your behavior, irrespective of your better judgment.
  "But I know someone else who did it."
  "I'll go along with what the group thinks."

- Allowing pride, ego and emotion to get in the way of good judgment.

- Being complacent.
  "I checked the fuel last night."
  "The weather at this time of the year is so predictable."
  "I'm too experienced for that."

- Being intolerant and impatient.

- Succumbing to "get-thereitis"
  (i.e. trying to complete a flight as planned irrespective of the risks).

- Being arrogant.

- Being overconfident in your abilities
  (i.e. having a false sense of security).

- Disregarding the airplane performance envelope and limits.

Pilots exhibiting these attitudes and behaviors take unnecessary risks and generally "get away with it" for a time, which simply reinforces their approach. However, statistics show that with such attitudes, the risk of making errors increases, thus increasing the probability of having an accident. It is, therefore, important for pilots to counter any hazardous attitudes and develop safe attitudes and safe behaviors.

Countering Hazardous Attitudes

It must be recognized that experience and skill cannot necessarily counter accident proneness brought about by hazardous attitudes. Furthermore, some hazardous characteristics are considered quite normal (even encouraged) in a non-flying environment. Pilots therefore need to:

> ▶ Be aware of any hazardous attitudes or behavioral traits they may have.
> ▶ Counter any hazardous attitudes.
> ▶ Adopt safe attitudes and behaviors.

An awareness of your personality, in terms of what hazardous attitudes you may have, can be obtained by completing the FAA self-assessment inventory. This can be found in the FAA Accident Prevention Program publication FAA-P-8740-53, Introduction to Pilot Judgment.

The next step is to continually examine your own thinking and, when a hazardous thought is recognized, counter that thought by an appropriate antidote thought from the following antidotes spelled out by the FAA:

Hazardous Attitude	Antidote
Anti-authority	Follow the rules. They are based on experience and are usually right.
Impulsiveness	Not so fast. Think before acting.
Invulnerability	It could happen to me.
Macho	Taking chances is simply foolish.
Resignation	I'm not helpless. I can make a difference. I am not going to stop trying.

The antidotes should be memorized for each of the hazardous attitudes so that they automatically come to mind when needed.

Self-awareness and the ability to counter hazardous thoughts will lead to safe flying behaviors. Examples of such behaviors include:

> ▸ Having a willingness and enthusiasm to learn.
> ▸ Being organized, systematic, thorough and disciplined.
> ▸ Being cautious, patient and unhurried.
> ▸ Being on guard for any errors or situations where errors might occur.
> ▸ Having the ability to exercise adequate controls over impulses, emotions and pressures.
> ▸ Being realistically aware of your own abilities and limitations.
> ▸ Being self critical and open to criticism.

Having the right attitude is the overriding factor that enables the deployment of all other risk reduction strategies. A realistic and honest self-assessment of your mental makeup is a good starting point.

Making Safe Judgments and Decisions

Courtesy of Australian Transport Safety Bureau

"Poor decision making processes was one of the factors that may have contributed to this accident"

The ability to make safe judgments and decisions, about both expected and unexpected events or problems, is essential for safe flying. The root cause of many pilot-related accidents has been traced back to errors in judgments and decisions, made before and during flight by both low and high time pilots.

Good pilot judgment can be defined as the ability to see and choose between the available alternatives, such that the choice made results in the safest practicable outcome. Judgment decisions generally involve a problem or choice, an unknown element, possible ambiguity, an evaluation of risk, usually a time constraint and generally stress. Not an easy process, but equally as important as the ability to takeoff and land.

Flying is dynamic and involves a constant series of judgments and decisions with one decision often impacting on another. One poor decision can increase the probability that another will follow and, as the chain of poor decisions increases, the alternatives for continued safe flight decrease.

Factors Inhibiting Good Pilot Judgment and Decision Making

An important starting point in learning good judgment and decision making, is to be aware of some of the factors that inhibit the process. These include:

Misunderstanding the Influence of Attitude

Attitude significantly influences a pilot's ability to make safe judgments and decisions. Hazardous attitudes can easily cloud a person's judgment. For example, an anti-authority attitude can lead to operating with insufficient fuel margin or descending below the minimum safe altitude. Alternatively, a macho attitude can lead to flying outside the envelope in an attempt to show off. Invulnerability can result in scud running at dangerous altitudes and conditions, particularly if you have got away with it before. The influence of attitude cannot be over-emphasized.

Failure to Appreciate the Impact of Stress

Stress has a significant influence on judgment. Some stress is beneficial, but when too much stress occurs, performance declines and judgment deteriorates. Pilots under too much stress (either life induced stress or aviation induced stress) are likely to take greater risks than they would otherwise. For example, a pilot under pressure and faced with deteriorating weather may decide to press on rather than turn back, preferring to take a chance rather than be inconvenienced.

Not Controlling Expectancy

Expectancy is when you are so focused on the objective (e.g. landing) that you ignore any warning signs of a gathering unsafe situation.

Insufficient Knowledge and Experience

A lack of knowledge and experience results in the pilot having an inadequate foundation for making sound judgments and decisions. For example, knowledge is needed for effective problem recognition as well as problem resolution. Past experience helps a pilot to identify what information is relevant and what is not. Inadequate knowledge and experience can result in misconceptions leading to a chain of inappropriate judgments and decisions.

Failure to Manage Pressures

Self induced pressure, or pressure from family or friends, can result in judgment being clouded, leading to potentially unsafe decisions being made. Avoiding any notion of "get-thereitis" will avoid such an issue. Under time pressures, pilots generally look for the first workable solution. That is, not all options are necessarily mentally reviewed. The more thorough the preflight preparation, the better your ability to make rapid judgments and decisions, as many of the options would have already been thought through in less stressful circumstances.

Not Monitoring Decision Feedback

A decision involves an action or an inaction that will have some form of feedback. If feedback is not monitored the subsequent decision(s) may be inappropriate.

False Hypothesis

There is a human tendency to rationalize things irrespective of flawed logic, or to make assumptions using incomplete information. All too often we "see what we want to see", "hear what we want to hear" and "do what we want to do". This is particularly the case when there is ambiguity and uncertainty. The danger is that decisions or actions may be based on what the pilot would like the circumstances to be, rather than what they are in reality. For example, lost pilots will sometimes try to guess where they are, then look for features that support their assumption and ignore inconsistent features. Situations in which false hypotheses are particularly likely to occur are:

- When expectancy is high.
- When attention is diverted.
- After a time of high concentration.
- When the false hypothesis acts as a defense mechanism
  (e.g. when you don't want to admit to a mistake).

Failure to Maintain Situational Awareness During Flight

In order to be able to address a problem during flight, it must first be detected. Failure to maintain situational awareness may result in the problem not being detected at the earliest opportunity, thus reducing the judgment and decision options. Furthermore, once a problem has been detected, focusing on the problem at the expense of maintaining situational awareness, can lead to other problems not being identified.

Getting Behind the Airplane

Not staying ahead of the airplane results in events and situations controlling the pilot's actions, rather than the other way round.

Ways to Improve Judgment and Decision Making

Your ability to make safe judgments and decisions depends upon your skill, recallable knowledge and experience. All three areas can be improved through training, education and gaining meaningful experiences. While the airlines have adopted Crew Resource Management (CRM) as an integral part of their pilot training programs, the general aviation pilot is normally not exposed to such material. Because judgment and decision making is such a significant contributor to accidents, most private pilots would benefit from some form of education on the subject. Ways to improve your judgment and decision making are to:

- Consciously take steps to avoid those things that inhibit safe judgment and decision making.
- Learn about the risks associated with flying and utilize strategies and tactics that minimize those risks. Understanding the risks and how to manage those risks, gives you a much better ability to choose and evaluate alternative courses of action in a timely manner.
- Have a "what if" approach and pre-plan your options during preflight preparation, when there are (or should be) no time pressures. Weigh up probabilities and always have options. This will reduce the possibility of stressful in-flight surprises.
- Ensure you have sufficient recallable knowledge and experience with an ability to translate that knowledge and experience into good judgment. Actions that can be taken to help this process include:
    - Practice the detection and management of problems such as failed equipment (e.g. alternator), failed instrument(s), the onset of engine failure, abnormal engine indications (e.g. high CHT), uncertainty in your position and so on. Then, if such problems are encountered, they will be familiar to you and, can be addressed rapidly if time is limited. Extensive practice also helps in translating recallable knowledge and skills into unfamiliar situations.
    - Gain new experiences by flying with a more experienced pilot.
    - Practice judgment and decision making by simulating various scenarios while flying with an instructor. Direct experience is invaluable.
    - After every flight, honestly review your judgments and decisions and determine if they could have been improved. Look at your mistakes and consider them a learning experience.
    - Take recurrent training to sharpen and extend your skills, as well as increase your confidence. Get the instructor to simulate scenarios that require judgment and decision making in a realistic flying environment.
    - Learn from the experience of others (both good and bad - but make sure you understand what is bad and what could have been done differently).
    - Analyze accident reports to identify the judgment and decision errors. In particular, what actions/decisions could have been taken to avoid the accident.
    - Be aware of your own limitations.
- Use a systematic problem solving process for handling in-flight problems, such as the following:
    - Look for changes or problems that can threaten safety. Maintain situational awareness so that a change or problem can be detected at the earliest opportunity.

> ▸ Ascertain the significance of the change or problem by gathering more information. Obtain outside assistance where practicable.
>
> ▸ Identify and assess the available alternatives. In some cases more information may be needed in order to make a rational judgment.
>
> ▸ Based on the judgments made about the options, decide on what action needs to be taken to eliminate and/or manage the risk. An important element is to recognize when a decision has to be made.
>
> ▸ Maintain constant vigilance to determine the effect of the decision, so that subsequent courses of action are not based on erroneous information or inappropriate expectations. Ensure you are not preoccupied or stressed.
>
> ▸ Expect the unexpected.
>
> ▸ Be prepared to deal with the situation as it is, and not as you planned it would be.

Safe judgment and decision making is a capability that can be developed and improved. Poor judgments and decisions can be fatal.

Knowing When Not to Fly

Flying is not an activity that can be undertaken safely any time you please. It is dependent upon your health, state of mind, qualifications, proficiency, weather, airplane capability, external pressures etc. These are all risk factors that should be assessed, and judgments made as to whether the options available pose an unacceptable risk. If they do, the flight should be either postponed or cancelled.

The following are examples of situations where the risks are, in all likelihood, unacceptable:

> ▸ When you are subject to certain physical, mental and physiological factors such as being:
>   - ▸ Unprepared.
>   - ▸ Worried or under severe stress.
>   - ▸ Fatigued.
>   - ▸ Unwell.
>   - ▸ On medication.
>   - ▸ Overconfident.
>   - ▸ Under pressure to get home.
>   - ▸ Frustrated or angry.
>   - ▸ Hung over.
>   - ▸ Unsure.
>   - ▸ Hungry.
> ▸ When you have not had time to adequately prepare for the flight.
> ▸ When the weather is marginal relative to your experience, skill, weather understanding and escape options.
> ▸ When your proficiency is below what is judged to be required for the flight.
> ▸ When you have pressures that impact on your ability to operate safely.
> ▸ When the airplane performance and serviceability does not fully meet the requirements of the flight.

The state of a pilot's health can, in particular, have a major impact on his or her ability to manage aviation risk. Even a minor illness can seriously downgrade performance of piloting tasks. Many medications (both prescribed and over the counter) have effects that impair judgment, memory, alertness, co-ordination, vision and other critical functions. Fatigue continues to be one of the most dangerous hazards to flight safety, as it may not be apparent to the pilot until serious errors have been made. Also stress, from various causes, can impair pilot performance, often in subtle ways. These and other health effects are discussed in more detail in Section 13.

Before every flight, your level of fitness can be checked using the acronym "I'M SAFE".

I	Illness	Do I have any symptoms?
M	Medication	Have I been taking prescription or over the counter medications?
S	Stress	Am I under pressure or unduly worried?
A	Alcohol	Do I comply with the regulatory alcohol and drug restrictions?
F	Fatigue	Am I tired and not adequately rested?
E	Eating	Am I adequately nourished?

The above criteria all relate to one or more of the four main areas of risk; namely the pilot, the airplane, the environment and external pressures. Whilst these criteria in themselves are relatively simple, the go/no-go decision is not always straightforward and it takes self-discipline, the right attitude and sensible judgment to make a safe decision.

In other words, it is all about managing risk. For most private pilots, this is left up to the individual, as there are generally no standard operating procedures to provide formulated guidance. However, the use of a Personal Minimums Checklist does provide the private pilot with an excellent framework for carrying out the preflight risk assessment process. The checklist, developed by the FAA, provides a simple mechanism for setting limits to risk factors in the four main areas of risk. It is a practical and flexible risk management decision tool that all private pilots can readily use. A copy of the FAA Personal Minimums Checklist, supplemented by practical methodology for setting various minimums, can be downloaded from: http://www.faa.gov.

It is unlikely that you will necessarily have all the risk factors identified, as these vary depending on the circumstances. A simple additional check therefore, is to ask three questions that were posed by R. Scott Puddy in his AVweb article *The Go/No-go Decision:*

> ▸   Are there any safety issues with the planned operation?
> ▸   What are the available options?
> ▸   Of these options, which are completely safe?

If there are no satisfactory options, then the flight should be abandoned.

Remember - "When in doubt, wait it out".

Being Well Prepared

3

Being thoroughly prepared is, in many ways, the key to safe flying. But, being prepared means far more than simply preparing a flight plan. It means:

> ▸ Examining the potential risks and hazards.
> ▸ Understanding the conditions likely to be encountered and the various factors that could affect the takeoff, en-route and landing phases of the flight.
> ▸ Applying appropriate risk minimization strategies.
> ▸ Carefully evaluating and pre-planning options.
> ▸ Ensuring that both you and the airplane are in a safe condition.
> ▸ Making certain you have the required knowledge and skills for flying the airplane and handling emergencies.
> ▸ Being mentally prepared for all possible eventualities.

Thorough preflight preparation significantly reduces the likelihood of making errors and improves your ability to make safe judgments and decisions. It is a lifesaver.

Having Sufficient Flying Knowledge

Adequate knowledge of all relevant (or potentially relevant) elements of a particular flight is an important pre-requisite to safe flying. There are many aspects of flying about which a pilot needs recallable knowledge and understanding. They can be categorized as follows:

> ▸ Self-knowledge (e.g. health, behavioral characteristics).
> ▸ Airplane knowledge (e.g. systems, performance, limitations, normal operating procedures, emergency procedures).
> ▸ Operational knowledge (e.g. operational procedures, regulations, weather, risk avoidance strategies).

Knowledge required for safe flying is extensive, and lack of knowledge has been demonstrated to be a contributing factor in many aviation accidents. Not only do you need basic knowledge about flying and operating an airplane, but you also need to be familiar with flying risks and the corresponding avoidance/minimization strategies and procedures. The broader your knowledge, the better your chances are of making a safe judgment or decision.

Ensuring you are adequately prepared for a flight in terms of knowledge, requires an appreciation that:

> ▸ Flying knowledge is obtained through a combination of training, education and experience.
>
> ▸ The extent of knowledge required for a particular flight depends on the nature of the flight and the possible eventuating circumstances (e.g. weather changes, instrument failure, ATC changes, engine failure). Deficiencies in that knowledge can result in inappropriate decisions, incorrect actions, distraction and additional stress.
>
> ▸ It is not enough to just memorize facts. It is necessary to understand the "why" of a topic or procedure to better help retain it in your memory, as well as having a greater chance of dealing with new and unique situations.
>
> ▸ Knowledge retention by human beings decreases with time, therefore regular "re-learning" is required.
>
> ▸ It is important to be aware of what you don't know.
>
> ▸ Flying knowledge on its own is not sufficient for safe flight. It must be coupled with skills, proficiency and experience.
>
> ▸ Knowledge puts you in a much better position to cope, if something goes wrong.

It is therefore vital to be up to date with all relevant aspects of flying knowledge and to be aware of one's knowledge limitations. Having a regular pattern of knowledge gathering and refreshment is recommended. For example:

> ▸ Plan your flight over several days and refresh your memory on relevant matters as they arise.
>
> ▸ Set up a program, over a period of time, to re-read the Pilot Operating Handbook (POH) and relevant aviation publications.
>
> ▸ Go through the regulations and highlight key aspects that are applicable to your flying. They can then be quickly reviewed from time to time.
>
> ▸ Read books and articles on flying.
>
> ▸ Attend safety seminars.
>
> ▸ Go through the emergency procedures prior to every flight.

A checklist of airplane knowledge needs is given in Appendix 1. This is a particularly important knowledge area, as there is a lack of standardization from one airplane to another, as well as an increasing complexity as you move to higher performing airplanes.

As many people have observed – there is safety in knowledge.

Being Proficient

In order to fly an airplane safely you must be proficient at carrying out all the required tasks. The level of proficiency required depends on the complexity and demands of the flight in question. In order to undertake a flight safely your proficiency must match the proficiency needs of the particular flight. For example, the proficiency needs when flying an approach in a high performance twin in bad IMC are significantly greater than when flying a small single on a sunny, clear and windless day. If you have not flown for some time, your proficiency will more than likely not be adequate. Even if you have flown recently, not all skills are necessarily exercised and hence your proficiency may have some limitations. When a pilot's proficiency does not meet the needs, or potential needs of the flight, there is an increased likelihood of making errors that in turn could lead to an increased level of risk.

The Meaning of Proficiency

Proficiency means having the competencies to adequately perform all the actions necessary to fly an airplane safely in the eventuating conditions. It should not be confused with currency. While the regulators have established flight review and recent experience criteria, these by themselves do not mean a pilot is proficient. Being "legal" is not always safe.

Proficiency requires:

> ▸ Experience.
> ▸ Currency (recent experience).
> ▸ Knowledge.
> ▸ Skills.

Flying is not like riding a bike. Proficiency in flying requires regular practice, as it decreases with time if the required skills and knowledge are not maintained. While physical skills (the ability to control an airplane) generally deteriorate slowly, cognitive skills (decision making and problem solving) deteriorate more rapidly over time. This should be factored into any proficiency maintenance program.

Assessing Your Proficiency

The private pilot is normally the only judge with regard to the level of proficiency required for the particular flight. Hence it is important that a realistic self-assessment be made. In doing so, it should be remembered that studies have shown that pilots have difficulty in assessing their own level of proficiency. Therefore, if there is some uncertainty, go for a check flight with an instructor to pin point any weaknesses. No matter what the nature of your flying is, the mandatory recent experience requirements may not be sufficient to always maintain your proficiency. Therefore, there is a need to develop a personal recurrent training program that matches your particular ability and the needs of your flying.

A self-assessment can best be performed by identifying the likely demands of the flight and asking yourself a series of questions such as:

> ▸ Is my level of skill and experience appropriate for the possible conditions?
> ▸ Can I recall the emergency procedures for the airplane to be flown?
> ▸ Can I recall the in-flight checks?
> ▸ Can I recall all the airplane performance parameters?
> ▸ Do I recall how to operate all the avionics?
> ▸ Have I flown sufficient hours in the last 6 months?

Honest answers to such questions will provide the basis for a safe assessment. If your proficiency does not match the needs of the flight, take remedial action.

Maintaining Proficiency

Proficiency can be maintained and enhanced in a number of ways depending on the extent and nature of your flying.

In order to maintain your proficiency, the first action is to always be aware of those things about which you need to have knowledge (e.g. airplane systems, operating procedures, emergency procedures, ATC procedures, risk reduction procedures). Brush up regularly on those things you need to know for your flying. Spend a sensible amount of time planning your flight and re-learning those things that are applicable (it is amazing how much there is to know and how quickly you forget). Use the preflight planning checklist in Appendix 2 to help ensure all relevant issues have been addressed.

The next step is to have an appropriate recurrent training program. If you fly infrequently, then it is best to plan your flying activities to maximize the frequency rather than the total time you spend flying. When flying, practice infrequently used skills such as cross wind landings, short field takeoffs and power off glide approaches. It is also wise to obtain a periodic flight check to identify weaknesses and refresh critical skills. Another option, is to practice maneuvers with another pilot or practice error inducing situations with an instructor. Infrequent flyers should also consider setting more conservative personal minimums by:

> ▸ Allowing greater margins for such things as fuel reserves, takeoff and landing distances, allowable cross winds and weather minima.
> ▸ Allowing more time for preflight planning and preflight inspection.
> ▸ Establishing conservative go/no-go criteria with respect to external pressures, airplane knowledge, weather and proficiency.
> ▸ Setting higher recent experience criteria compared with the minimum legal requirements (e.g. number of takeoffs and landings, total hours in the airplane type, number of approaches).

Some skills can also be refreshed through the use of a personal computer-based aviation training device (PCATD). You can practice instrument scanning, instrument flying, IFR approaches and emergency situations such as engine failure and instrument malfunction. You can also practice a planned flight by simulating as many of the activities as possible. Go through all the motions of a normal flight (in real time) including preflight planning, all checks, radio calls (say them out loud), fuel monitoring, NAV/COM changes, fuel tank changes, instrument monitoring and so on. The process can also be helped, by first writing down the sequence of activities from engine start to engine shutdown, noting such things as frequencies to be used, all likely radio calls, altitude changes and navigation requirements. The whole exercise has the added benefit of mentally preparing you for the flight as well as identifying any knowledge gaps.

Part of your proficiency program should also include taking every opportunity to enhance your flying experiences. Some suggestions include:

▸ Acting as co-pilot for a more experienced pilot or inviting another pilot to accompany you and asking them to observe and make comments.

▸ Talking to other pilots and learning from their experiences. However, learn to distinguish worthwhile experiences from less appropriate practices.

▸ Sitting in the back seat of an airplane and witnessing other pilots flying dual.

▸ Making every flight a learning experience. For example, maintain a log of errors and how best to avoid them in the future. Review the list as part of your preflight preparation.

Note that if you are proficient on one type of airplane and then want to fly a different type, which you have not flown for some time, you cannot consider yourself proficient in the latter airplane. Suitable dual instruction and practice may be required in addition to refreshing your knowledge of the POH. A lack of proficiency and flying an unfamiliar airplane, is a deadly combination.

These suggestions will vary with the individual pilots, but illustrates the need for a program of activities that will help maintain a level of proficiency that matches the needs of your flying. Importantly, avoid becoming complacent or over-confident.

Thorough Preflight Planning

Preflight planning is one of the major safeguards against avoidable errors. It provides the basis for a safe flight and should never be underestimated. Analysis of accidents has revealed that preflight planning is often inadequate or entirely ignored.

Preflight planning involves gathering information, evaluating that information and making a series of judgments and decisions about the priorities, objectives and requirements of the flight. It includes making a rational assessment of the risks of all relevant options and making certain that the decisions reflect safety as the number one objective of the flight. And, it provides the information that will be needed during the flight to achieve the desired outcome in the safest possible manner.

It is important that sufficient time is allowed for preflight planning and that this is done in a thorough and unhurried manner, as errors made during the planning phase can have dire consequences. Flight planning takes time. A good idea is to start planning your flight several days ahead – but don't let this lock you into a poor decision to go. Furthermore:

> ▸ Become familiar with all the relevant information that pertains to the flight and use all the resources that are available (e.g. people, the Internet, manuals).
>
> ▸ Go through the entire flight in your mind to help identify planning needs and knowledge refreshment requirements.
>
> ▸ Carefully review the potential hazards and plan the flight taking into account any relevant risk reduction strategies.
>
> ▸ Allow adequate margins of safety and set personal minimums commensurate with your level of knowledge, experience and proficiency.
>
> ▸ Anticipate as many things as you can before leaving the ground. Run through "what if" scenarios and what action could be taken to be ready for variation from the expected (e.g. pre-planning possible escape routes should the weather deteriorate).
>
> ▸ Night flying has additional hazards, which requires care when planning the flight. The weather requires particular close attention, as visual clues of deteriorating conditions are not necessarily available in the dark. When planning your escape routes take into account the availability of airports, with runway lighting, along the intended route.
>
> ▸ Check the flight plan for completeness and accuracy. Accuracy should be checked by applying mental logic to headings, ground speeds and drift.

Preflight planning is about risk minimization and ensuring that safety is always the number one priority of the flight. In other words, preflight planning provides the basis for judgment and decision making both before and during the flight. By thinking through all the issues and options beforehand, you are much better prepared for any eventuality, either planned or unplanned. It allows you to take all the actions necessary to minimize any potential risks in an unhurried and calm environment. Even a few circuits at the local aerodrome require some preflight planning. You can, for example, experience communications failure, have an engine failure on takeoff, encounter a shift in the wind direction necessitating a crosswind landing, and so on. The point is, that every situation is different depending on the pilot, the aircraft, the environment and the external pressures. Allowing plenty of time for preflight planning, and taking advantage of that time, is fundamental to safe flying.

Preflight Planning Checklist

In order to help the planning process, a detailed checklist of most of the issues that need to be considered is given in Appendix 2 and covers the following key elements:

▸ Personal preparedness/readiness.
▸ Personal minimums.
▸ The go/no-go decision.
▸ Passengers.
▸ Weather.
▸ Weight and balance.
▸ Airplane serviceability.
▸ Departure.
▸ En-route.
▸ Destination.
▸ Personal equipment, data and information.
▸ Flight plan.

In addition, several aspects of preflight planning are discussed elsewhere in this manual, in more detail, as follows:

▸ Fuel preflight planning – Section 5.
▸ Weather preflight planning – Section 6.
▸ Airframe icing avoidance preflight planning – Section 7.
▸ Preflight planning that reduces the risks in the event of an engine failure – Section 14.

Familiarity with the checklist and the specific details in relevant chapters will help to achieve the safest practicable outcome. Never ignore, or treat lightly, the importance of unhurried and thorough preflight preparation and planning.

Thorough Preflight Airplane Inspection

The importance of a thorough preflight airplane inspection cannot be over-emphasized. A cursory and hurried inspection is of little value. You need to know why a particular item is being checked so that you are actually looking for discrepancies, rather than just going through the motions.

The preflight inspection checklist in the POH should always be followed, as there are design differences between airplanes. Also, particular items may need special attention. Appendix 3 provides a comprehensive preflight inspection checklist for a typical single engine airplane that can be used to augment the specific airplane POH checklist where relevant.

Prior to inspecting the airplane, ensure you are familiar with all the airplane features (e.g. know where all the fuel drain points are located). In addition:

> ▸ Be prepared to delay or cancel the flight if an unacceptable discrepancy is found.
> ▸ Never take another person's word that the airplane is ready and/or has been refueled.
> ▸ While it is necessary to check the fuel for contamination after refueling and before the first flight of the day, some pilots check the fuel before every flight.
> ▸ Fully preflight check your airplane at intermediate stops. (While not every element needs checking it is a good habit and ensures nothing is missed.)

Never fly an airplane that you are not completely satisfied with.

Identifying Problems with the Airplane

Apart from visually checking the airplane, you also need to use your other senses (touch, smell and hearing) to help identify possible problems.

> ▸ Listen for strange noises (e.g. grinding, scraping, slipping, cable slapping and knocking sounds).
> ▸ Feel for excessive play or restrictions in all moveable controls.
> ▸ Determine the smoothness of control surface movements and check for binding or higher than normal friction.
> ▸ Check the security of controls (e.g. trim tabs – an important check to avoid flutter).
> ▸ Feel for nicks and dents in the propeller(s).
> ▸ Gently wriggle hinge parts to check for excessive free play.
> ▸ Gently push and pull connecting rods to check for play or looseness.
> ▸ Smell for fuel leaks.

Possible Problems

The process of inspection is also helped, if you have some appreciation of what types of problems might be uncovered. The following are examples of things that have been found:

> ▸ Cables incorrectly routed after maintenance. (It has happened more than once!)
> ▸ Wear in Heim joints (ball joint) at the end of push rods.
> ▸ Insects in the pitot tube or fuel tank vents.
> ▸ A bird nest in the engine compartment.
> ▸ Nicks in the propeller(s).
> ▸ Incorrect tension in the control cables.
> ▸ Missing rivets.
> ▸ Tires with insufficient pressure.
> ▸ Heat affected electrical/instrument wiring.
> ▸ Failed/worn bearings in aileron pulleys.
> ▸ Bent under-wing vent tube.
> ▸ Missing hinge safety wires.
> ▸ Corrosion.
> ▸ Hinge split pins missing.
> ▸ Bald tires or tires with localized wear.
> ▸ Loose aerials or lights.
> ▸ Leaking fuel.
> ▸ Leaking hydraulic fluid.
> ▸ Locking wire missing from nuts.
> ▸ Control cables contacting part of the airplane.
> ▸ Worn control cables.
> ▸ Loose nuts.
> ▸ Excessive control surface hinge play.

Taking the time to do a thorough preflight inspection is essential for safe flying.

Being Mentally Prepared

Flying is not always predictable and things often need to be done with little delay. You therefore need to be mentally prepared for the unexpected as well as the expected. For example:

> ▸ Weather can change, requiring a diversion or return.
>
> ▸ Headwinds can be stronger than forecast with fuel consumption increasing accordingly.
>
> ▸ The vacuum pump can fail, with the result that certain instruments are unavailable.
>
> ▸ The engine can fail, requiring execution of an engine off landing (rare but a possibility).
>
> ▸ Carburetor icing can be encountered, necessitating appropriate remedial action.
>
> ▸ The landmark you were anticipating does not appear, requiring decisions on what actions to take.
>
> ▸ Departure may have to be delayed.
>
> ▸ An instrument can malfunction or an electrical failure can occur, requiring decisions and action to avoid a worsening situation.
>
> ▸ A door can pop open, or the engine may sound abnormal.

And so on.

Being mentally prepared for such possibilities and knowing how to deal with them avoids becoming stressed and making additional errors. In other words, you are mentally prepared to accept that such events are possible, you have the willingness to act and have the ability to achieve a safe outcome. Flying, is therefore, as much about being mentally prepared as it is about being physically prepared.

Avoiding "Cockpit" Errors

4

How we manage situations in the cockpit can significantly influence our ability to avoid errors. For example, failure to detect that the carburetor heat is not functioning can lead to possible loss of engine power, if conditions are conducive to carburetor icing. Becoming distracted, and neglecting to lower the landing gear can result in a gear-up landing. Allowing too many tasks to accumulate can cause unnecessary errors to be made, because of cockpit overload (i.e. getting "behind the airplane"). Not concentrating, or not listening properly, to the radio can result in communication breakdown, with the consequent possibility of incurring a collision (either on the ground or in the air). Spending too much time with your head down can easily result in loss of situational awareness, and becoming lost. All these errors can be defined as "cockpit" errors, which in some cases, can lead to further errors. However, such errors can *readily be minimized* by a disciplined approach to checks, staying ahead of the airplane, avoiding distractions, avoiding communication breakdown and maintaining situational awareness.

A Disciplined Approach to Checks

Checks provide a structured means of breaking a possible error chain at the earliest opportunity. They take two basic forms – formal and informal checks. Formal checks, by way of checklists, are designed to provide the safety margin needed to counter the human tendency of overlooking a vital action or detection of a fault, when distracted, under pressure, fatigued or simply complacent. In addition, a pilot's memory is not always perfect. Informal checks, on the other hand, are carried out on an 'as needs' basis and involves ascertaining the correctness of an action, checking the perception or understanding of an issue, or checking the settings and functionality of a system. They can range from confirming a clearance, to checking that the fuel switch is in the correct position, or determining the status of an instrument.

Checklists

While it is preferable to always use a written checklist, this is not always practical when there is only one pilot. In single-pilot operations, it is, therefore, generally recommended that:

> ▶ Written checklists be used for all ground operations to avoid inadvertently overlooking or forgetting an important safety item or sequence.
> ▶ All emergency checks be committed to memory.
> ▶ Normal in-flight operational checks be committed to memory.

Acronyms can be used as a technique for remembering in-flight checks provided they are brief and relatively simple. It is particularly important to practice in-flight checks, including emergency checks, as they will be readily recalled when required.

All airplanes are different, and even the same make and model can have differences because of subsequent modifications. Therefore, it is necessary that checklists from the specific airplane POH be used.

When using checklists, it is important not to simply go through the motions, but to actually understand the purpose behind the check and its relevance to safety. A mechanistic approach to checklists leads to errors of expectation. Checklists should in fact be regarded as cue cards that prompt you to:

> ▶ Challenge yourself, by asking a question (generally the left hand side of the checklist item).
> ▶ Physically check the status or condition of the item, about which the question is being asked.
> ▶ Respond to the challenge, by saying out loud what was physically checked.
> ▶ Check that the response is correct (i.e. the answer matches the right hand side of the checklist).

For example, a checklist item will read: MIXTURE................ RICH
The process of carrying out the check would be as follows:

Challenge:	"Mixture?"
	Check the position of the mixture control and say what you see and feel.
Response:	"Full Rich"
	Check that this is the correct response.

It is also good practice to:

▸ Double-check the condition of critical items.
▸ Touch the item being checked.
▸ Avoid rushing checks.
▸ Complete all checks, even though the airplane may have recently been flown (e.g. a short refueling stopover).
▸ Go back to the beginning of the checklist, if you are interrupted.

Often pilots link a memorized check to a flight event or situation. While this may be a useful "trigger", a problem can arise if something happens in a non-standard way, or you become distracted.

A problem with some checklists is that they contain routine procedures, with the result that the important safety items can easily be lost in significance. John Deakin, in his AVweb article *Throw Away Those Stupid Checklists*, argues that written checklists, for cockpit checks, should be comprised of essential safety items only, as long written lists can lead to a mechanical approach without the mental awareness of what is being checked and why. Deakin suggests that initially, all items in the cockpit should be checked using a mental "flow pattern" (a logical flow through the cockpit) followed by the use of a simple written checklist to confirm the key safety checks.

Whichever checklist methodology is used, the key thing is to do them and to do them correctly.

Informal Checks

Flying involves numerous demands on our attention; with the result that it is possible to make mistakes or not notice that something is amiss. The human brain can also make assumptions with potentially dangerous results. Regular checking of all flight conditions and actions reduces the likelihood of such errors from developing, and is a necessary part of safe flying.

Staying Ahead of the Airplane

A high workload, or an unexpected event during a flight, can lead to stress, mistakes and loss of situational awareness. One way of reducing these possibilities is to always be thinking and planning an action ahead of when it is needed. That is, staying ahead of the airplane.

There are a number of ways of ensuring you stay ahead of the airplane.

By having adequate proficiency, you can convert certain tasks to more routine actions, needing minimal processing resources, which, in turn is less demanding. This results in less stress and therefore, allows your mind to more readily cope with other things. Ensuring you are thoroughly prepared before any flight reduces the mental processing requirements needed during the flight, thereby reducing the potential for cockpit overload and the likelihood of making errors. During flight, there are two guiding principles for staying ahead of the airplane – anticipate as much as you can, and be pro-active rather than reactive. Specifically:

> ▸ Think and plan ahead, while flying, by regularly monitoring your situation so that actions can be anticipated and planned. Anticipate such things as radio calls, frequency changes, altitude changes, fuel tank changes, course changes, upcoming landmarks or waypoints, position reports, the likely duty runway, in-flight checks, controller handover and ATC instructions. Set everything up in the cockpit, ahead of time, for the next phase of activity and anticipate and plan ahead for any potential alternatives that may arise.
>
> ▸ Always be on the lookout for warning signs of potential problems so that remedial action can be anticipated and taken at the earliest opportunity. Warning signs can include abnormal instrument readings, deteriorating weather, silence on the radio for some time, vectors different from that anticipated, engine rpm dropping or a drift off course. Any abnormality should be seen as a warning sign.
>
> ▸ Re-schedule head-down work to periods of low activity.
>
> ▸ Schedule or re-schedule activities so as to minimize task conflicts.
>
> ▸ Always expect the unexpected and be mentally prepared for an emergency.

Part of staying ahead of the airplane is not to lose sight of the routine activities. That is, a regular scan and cross check of the instruments and gauges, a regular check of the fuel, maintenance of a fuel log, a regular position check, a regular look outside of the cockpit and a continuous check on the radio communications. At the same time you need to guard against those situations that diminish your ability to stay ahead of the airplane, namely:

> ▸ Being pre-occupied.
>
> ▸ Accepting ambiguous information.
>
> ▸ Being distracted.
>
> ▸ Having too much head down work (e.g. recalculating headings using a hand held computer rather than making quick mental calculations using rules of thumb).
>
> ▸ Being fatigued.
>
> ▸ Being focused on one problem at the expense of other required actions.
>
> ▸ Being unprepared for instrument indications that are other than expected.
>
> ▸ Having not adequately pre-planned the flight.

Staying ahead of the airplane smoothes out the workload, avoids cockpit overload and reduces the risk of losing situational awareness. It is an extremely effective risk reduction strategy. The key is to use your spare time to think ahead, plan ahead and visualize your next situation.

Avoiding Distractions

Distractions can cause actions or responses to be missed, with potentially dangerous consequences. Diverted attention is particularly likely when the pilot is under time pressure or stress. Even a minor abnormality under these circumstances may distract a pilot from other aspects of the flight.

Distractions can be either pilot related or due to external factors. The main causes of distraction are:

> ▶ Too many concurrent tasks.
> ▶ Too much communication.
> ▶ Too much head down work.
> ▶ Responding to an abnormal situation.
> ▶ Pre-occupation with one task (to the detriment of other tasks).
> ▶ Interruptions.
> ▶ Having little to do and drifting into absentmindedness.

Fortunately, many distractions and interruptions can be avoided by taking some simple precautions that include:

> ▶ Keeping your mind on flying. There is always something to do and it will keep your mind from wandering.
> ▶ Staying ahead of the airplane so that you avoid being distracted by having too many concurrent tasks or being distracted by totally unexpected occurrences and situations.
> ▶ Using checklists and double-checking critical items.
> ▶ Being completely familiar with the airplane systems and procedures so that distractions will not lead you to apply inappropriate learning.
> ▶ Being systematic and methodical.
> ▶ Maintaining a sterile cockpit so non-essential conversation does not distract you.
> ▶ When tasks must be performed concurrently, avoiding letting attention focus too long on one task by setting up an appropriate scan.
> ▶ Ensuring that simple things don't distract you from flying the airplane (e.g. a dropped pencil).

If you have a qualified co-pilot, he or she should be encouraged to monitor the general performance and cockpit management, and advise of any anomalies, no matter how trivial they may seem.

However, not all distractions and interruptions can be avoided and it is therefore important to be able to recognize when you are vulnerable. The best way is to treat all distractions and interruptions as a warning signal. Extra vigilance can then be applied and action taken to avoid any potential errors. This can be helped, by planning how distractions are to be handled. For example, if interrupted while carrying out a series of checks, be ready to either start the checks again from the beginning, or mark the point at which the interruption occurred. Always ask yourself what you were doing before the interruption or distraction and decide what action is needed to get back on track.

Avoiding and countering distractions requires self-discipline and constant attentiveness.

Avoiding Communication Breakdown

While most miscommunication does not lead to serious consequences, it is an ever-present risk that has the potential to result in a dangerous incident or accident. Constant attention is therefore required.

Reasons for Communication Breakdown

There are many things that can lead to communication breakdown; an obvious cause is, simply, failure to communicate. Other causes are:

- Distraction, stress and workload.
  "I did not really hear what was said." (i.e. not listening)
  "I didn't check what was said."
- Expectation of an instruction with the result that a different message is heard.
  "I heard what I expected to hear."
- Ambiguous phraseology or unintelligible words used.
- Misheard or misunderstood instructions.
  "I thought that I heard............."
- Read back errors.
- Failure by ATC to hear any read back errors.
- Inattention.
- Clearance amendments not acknowledged by the pilot and not challenged by ATC.
- Not making timely or accurate reports.
- Radio not switched on.
- Volume turned down.
- Instructions issued too quickly.
- Incorrect squelch setting.
- Transmitting on the wrong frequency.
- Frequency not known.
- Loose headset jack.
- Out of range.
- Master avionics switch turned off.
- Speaker/headphone switch incorrectly positioned.
- Circuit breaker popped.
- Stuck microphone button.
- Misheard call sign.
- Incorrect audio panel switch position (a common error).
- Not checking NOTAM's for frequency changes.
- Poor use and understanding of English.
- Carrying out-of-date charts (with the wrong frequencies).

Ways to Avoid Communication Breakdown

Firstly, be aware of the many circumstances that can result in communication breakdown and guard against their occurrence. Many just require a simple check and attention to detail. Secondly, make sure you know how to use the equipment. Avionics vary considerably from one airplane to the next – a thorough briefing before climbing into a rented airplane is recommended. Some additional precautions include:

> - Keeping air traffic control informed of your intentions.
> - Lowering your voice pitch slightly.
> - Making sure the correct "transmit" radio selection has been made in dual radio airplanes.
> - Ensuring the correct frequency has been set.
> - Asking for something to be repeated if in doubt ("say again").
> - Asking the controller to "stand by" if you are not ready to receive communication.
> - Asking for clarification if unclear or uncertain about an instruction.
> - Transmitting in clear, concise and standard terminology.
> - Listening before speaking so that you don't over transmit or interrupt a transmission.
> - Thinking before speaking (rehearsing the transmission in your mind).
> - Listening carefully to every clearance.
> - Not having expectations of what you may hear.
> - Avoiding making assumptions.
> - Maintaining situational awareness so that communication can be crosschecked (e.g. having knowledge of terrain hazards and lowest safe altitude).

If there is no obvious cause of a communication failure then:

> - Try another radio.
> - Use the loud speaker and hand held microphone.
> - Request another airplane to relay your message.
> - Try plugging the headset into the passenger socket (if possible).

If still no success, follow the official communication failure procedure.

Maintaining Situational Awareness

Situational awareness is always being aware of where you are (such as position, terrain features, location of airfields), what is happening with the airplane (such as fuel consumption, engine operation, altitude, heading) and what is going on around you (such as weather, traffic, ATC requirements). Loss of situation awareness has been a contributing factor in numerous accidents and occurs when the pilot has an erroneous perception of the state of the airplane, the state of the external environment or the relationship between the two. This in turn can lead to a series of poor or flawed judgments and decisions, the outcome of which is an increase in the level of risk and the possibility of an accident.

Courtesy of Transport Canada Aviation Safety Newsletter – Redrawn by Liam van Deth

Causes of Loss of Situational Awareness

Situational awareness can be quickly lost if you are distracted or not ahead of the airplane. Other causes include:

> ▸ Complacency.
> ▸ A high workload.
> ▸ Too much head down time.
> ▸ Being caught off guard.
> ▸ Making false assumptions (e.g. reacting to an expectation rather than the true situation).
> ▸ A fixation on one problem, or issue, at the expense of all other necessary activities.
> ▸ Not being sufficiently fit to cope with the workload.
> ▸ Incomplete or poor communication.
> ▸ Not being prepared for the unexpected.
> ▸ Ambiguous information leading to confusion or inappropriate decisions.
> ▸ Using out of date procedures and/or charts.
> ▸ A lack of knowledge about the airplane, resulting in an inability to identify a problem and/or deal effectively with it.
> ▸ Over-reliance on the GPS (it can potentially fail).
> ▸ Distractions.
> ▸ Loss of communication.
> ▸ Fatigue.
> ▸ Improper procedures.

How to Improve Your Situational Awareness

Situational awareness (as defined by Crew Training International of Memphis Tennessee) is achieved by constantly searching for and collecting information about what has happened, what is happening and what is likely to happen. You can improve your situational awareness by avoiding those situations that cause loss of situational awareness, in particular, making sure you are not "behind" the airplane. In addition:

- ▸ Regularly evaluate your performance against the flight plan.
- ▸ Always maintain a state of vigilance and, if you feel something is wrong, take it as a warning sign and investigate.
- ▸ Avoid focusing on the detail for too long at the expense of the broader picture and visa versa. Divide your attention appropriately.
- ▸ Be prepared for possible unplanned events (e.g. bad weather, different ATC instructions). Have a "what if" approach with pre-planned options in mind.
- ▸ Listen carefully to all radio communications and visualize your position in relation to the ground and other airplanes (irrespective of whether you are in controlled airspace or not).
- ▸ Verify single source information wherever possible (e.g. looking for traffic when cleared for takeoff, confirming a GPS location by visual reference or VOR bearings).
- ▸ Gather data from every possible source, consider all the possible interpretations, don't jump to conclusions and don't make the data fit your hypothesis.
- ▸ When under pressure, step back and prioritize the actions.
- ▸ Stay focused, despite any distractions.
- ▸ Having embarked on a course of action (e.g. diverting), mentally stop every now and then and take stock of the situation.

If you lose situational awareness – firstly, fly the airplane, then assess the situation, evaluate the results and take positive action.

Avoiding Fuel Mismanagement

5

Fuel starvation (engine stoppage due to fuel interruption when there is still fuel available) and *fuel exhaustion* (engine stoppage due to depletion of all available fuel), have been major continuing contributors to airplane accidents.

"Fuel starvation was one of the factors that may have contributed to this accident"

Errors that have resulted in fuel related accidents include:

- Water contamination.
- Use of an incorrect fuel.
- Selection of the wrong fuel tank.
- Incorrect calculation of the fuel required.
- Inadvertently switching off a fuel tank.
- Higher than expected fuel consumption.
- Inadequate fuel monitoring.
- Incorrect operation of the fuel selector valve.
- Fuel vent blockage.
- Incorrect operation of the auxiliary feed pump.

Most of these errors can be attributed to pilot mismanagement and can therefore be prevented. Lessons from past accidents indicates that they can be easily avoided by:

- Pilots ensuring they have a sound understanding of the fuel system and the airplane fuel related performance parameters.
- Careful and prudent preflight fuel planning.
- Conducting a thorough preflight fuel check and inspection.
- Systematic in-flight fuel management.
- Following acknowledged safe fuel management practices.

Always take a conservative approach with regard to fuel and never take short cuts. Fuel exhaustion and starvation can be deadly.

Fuel System Understanding

An airplane should not be flown unless the pilot has a thorough understanding of the particular airplane fuel system and operating procedures as outlined in the POH. Note that it is a dangerous practice to apply a common set of procedures to fuel management, as there are many differences between airplane types. A lack of specific airplane knowledge can lead to an unsafe action, such as switching to the wrong tank. Fuel system understanding requires an appreciation of the overall fuel system and its operation, as well as knowledge of the following:

- Procedures relating to:
    - Refueling.
    - Fuel sampling and contamination checks.
    - Vent operation and inspection.
    - Sump/filter draining.
    - Mixture control (how and when).
    - Operation of the electric fuel pump (where fitted).
    - Fuel management and tank switching procedures.
    - Operation of all fuel selector valves, handles, pumps and gauges.
    - Requirements during takeoff and landing.
    - Operation of the auxiliary fuel system (where fitted).

> ▸ Fuel feed system and associated equipment, including electric fuel pump design and operation (where fitted).
> ▸ Number, type and arrangement of the fuel tanks.
> ▸ Usable fuel capacity in each tank.
> ▸ Grade of fuel required.
> ▸ Fuel venting system including any fuel filler cap vents.
> ▸ Location and function of all fuel drains and vents.
> ▸ Methods of prevention, detection and elimination of water in the airplane fuel system.
> ▸ Fuel consumption rates and the basis for these figures.
> ▸ How to determine the airplane range and endurance.
> ▸ Determination of the fuel state at any time during the flight.
> ▸ Methods of estimating required/available fuel resulting from diversions or wind.
> ▸ Any fuel tank design limitations.

In addition, you should be aware of any fuel system modifications that may have been made (e.g. long-range tanks) and ensure that the procedures reflect such modifications.

There are numerous examples where pilots have not thoroughly understood the fuel system, resulting in simple errors being made which in turn have led to an accident.

Preflight Fuel Planning

Careful preflight fuel planning can significantly reduce the risk of having a fuel related accident. Preflight planning provides an excellent opportunity for the pilot to review his or her level of understanding of the fuel system and pre-plan all the fuel related aspects of the flight, in a thorough and unhurried manner. Of particular importance is the fuel margin or contingency that is provided. Never plan to use fixed or variable reserve fuel.

Before you plan a flight, check the availability and operating hours of fuel services en-route so that appropriate refueling stops can be planned.

Usable Fuel Capacity

Fuel tank usable fuel capacities can be obtained from the POH. Filling the tanks to capacity will give you the maximum margin, but you will need to check:

> ▸ The weight and balance, as you may not be able to carry full fuel. Alternatively you may not be able to load all passengers or all baggage.
> ▸ The takeoff and landing limitations, as this may also limit fuel, passengers or luggage.

Whatever quantity you start with, it is crucial to know exactly how much usable fuel you have in the tanks.

Fuel Consumption Planning Rate

While fuel consumption rates are given in the POH, it must be recognized that these figures are based on optimally tuned new engines, operated under optimum conditions and correct mixture settings. As the result of service and wear, you can expect a fuel consumption figure up to 20% more than the published rate. Alternatively, you can use the average fuel consumption rate recorded over a period of time, but remember that the fuel consumption rate is dependent on altitude, power settings, temperature, mixture settings and airplane configuration. A suitable margin therefore needs to be added to the fuel consumption planning rate or the overall fuel quantity required.

Required Fuel Quantity

In calculating the fuel required, quantities need to be provided for:

- Start, taxi, takeoff, climb and departure.
- Cruise.
- Descent, approach and landing.
- Any required alternate.
- Any planned holding.
- Fuel reserve.
- Fuel margin.

The fuel reserve is that quantity of fuel that should normally remain in the airplane until after the final landing and is established on the basis of any regulatory requirements and your personal minimums.

A fuel margin or contingency is that quantity estimated to cover the following possibilities:

- Weather that is different from that used for planning purposes that can result in delays or the need for a diversion (e.g. higher headwinds, lower tailwinds, cloud, icing, fog, reduced visibility). Note that some pilots double any headwind, lower any tailwind and use the amended figures for the cruise fuel calculation. If the weather is marginal at the planned destination, make sure there is enough fuel to go to an alternate.
- Higher than normal fuel consumption due to changes in altitude, power settings, mixture settings or use of carburetor heat. Note that operating at high power can significantly increase fuel consumption.
- Unforeseen route changes, altitude changes, holding or maneuvering.
- Other unforeseen conditions.

The fuel margin or contingency that you provide depends on the nature of the flight and your personal minimums. But, in all cases some margin or contingency should be provided (nothing ever goes exactly to plan). Some pilots opt to always fill the tanks to capacity to maximize the margin.

If the fuel calculations show there is inadequate fuel for the flight, do not reduce your reserves or margins and don't proceed on the basis of relying on the gauges to see if you have enough fuel left. Safe options are:

> ▸ Change the flight plan to refuel at an intermediate point (check the availability of fuel en-route).
> ▸ Don't go.
> ▸ Use an airplane with a longer range.

Note that errors have been made in fuel calculations, and hence they should be carefully checked (in particular any conversion calculations). The results of electronic calculators should be checked using mental calculations or tables.

Fuel Endurance

It is always advisable to compute the fuel endurance for your airplane based on a conservative fuel consumption rate. Subtracting the time equivalent of your reserves and any personal minimums, will give you the available "time in your tanks".

Refueling and Preflight Fuel Checks

Courtesy of Air BP

A thorough preflight check of the fuel quantity, quality and system operation is vital before each flight and after refueling. Many accidents have resulted from improper and careless preflight fuel checks. The following guidelines will minimize the risk of errors occurring.

Refueling

Whenever possible, refuel the airplane yourself or observe the refueling to ensure the correct fuel is used and the correct tanks are being filled. Refuel on level ground but if this is not possible, take any unevenness into account. Try to leave tanks full overnight to minimize condensation.

Fuel Quantity on Board

Prior to any flight (including intermediate stops) a pilot should always physically check the fuel quantities in the tanks, making sure that they are either full, filled to the tabs or dipped with a properly calibrated dipstick. Any other visual check (e.g. judging the level relative to the tank wall) will not give an accurate result. In addition, there are some "Nevers":

- ▸ Never rely on the refueler to have filled the tanks during your absence.
- ▸ Never rely on a check made the night before the flight.
- ▸ Never rely on the word of another person.
- ▸ Never rely on the fuel gauges (they are notorious for their inaccuracy).
- ▸ Never rely on memory to recall the amount of fuel in each tank - write it down.

Before start up, note the fuel gauge readings before and after switching on the master, as a gauge needle may be stuck. Compare the gauge readings with the tank visual readings and write down the fuel quantity on board in the fuel log.

Fuel Contamination Check

Water in fuel can result from condensation in the tanks, from leakage past the filler cap or from water present in the supplied fuel. Appropriate sampling must therefore be carried out to ensure the fuel supply is free from water or other contamination.

- ▸ Always carry out a fuel contamination check taking a generous sample from the fuel tank drains and the filter before each flight, and after refueling. Free water will settle in the bottom of the sampling container while any suspended water will give the fuel a cloudy appearance.
- ▸ If a visual check for water is not conclusive, the only certain method is to use water sensitive paste or paper.
- ▸ Note that during refueling, the stirring action in the tanks puts any water into suspension. Therefore, it is advisable to take a sample before, as well as after refueling. Before taking a sample after refueling, wait for any water to settle and make sure the sample is of sufficient quantity to conclusively ensure no water is left. (Note that settling can take up to 15 minutes.)

After checking for contamination (e.g. water, dirt), confirm that the drain valves are properly closed and not leaking.

Fuel Type

Confirm that you are using the correct grade of fuel as recommended by the engine manufacturer by checking the color of the fuel sample taken for the contamination check, although the reliability of this check depends on the quantity of fuel taken on board.

Fuel System Equipment

The preflight inspection must include a careful check of the fuel system condition and operability.

> - Check that fuel tank vents are not blocked. A blocked vent can eventually result in inhibited fuel flow to the engine.
> - Ensure that all drain cocks and fuel tank caps are properly closed and sealed prior to flight. (Don't fly if there are any fuel leaks.)
> - Check the condition of the fuel cap seal.
> - Check for fuel stains or leaks during the walk around and when checking the fuel cap over the wing.
> - After engine start-up and during the taxi run, check the operability of the fuel tank feed system in accordance with the POH.
> - Always check that the electric fuel pump is functioning properly (where fitted).
> - Check the fuel tank lever detents to ensure the lever correctly stays in position.

It is a good idea to double-check all fuel related activities to avoid any chance of error. Never take short cuts where fuel is concerned. Fuel contamination, exhaustion and starvation can be deadly.

After-Start Fuel Management

A systematic approach to after-start fuel management provides a sound basis for avoiding fuel exhaustion or fuel starvation. As a precursor, review your knowledge of the following during preflight planning:

> - How to estimate the fuel needed at any time during the flight.
> - How to estimate the fuel required for any diversions and/or wind changes.
> - The power settings for best range and endurance.

Before Takeoff

The POH procedures for fuel tank selection and fuel flow checks vary. Some procedures suggest selecting the fullest tank during run up while other procedures are not that clear. However, the time from run up to takeoff can sometimes be quite short, leaving insufficient time for a proper fuel flow check. A better option is to select the fullest tank prior to engine start. Sufficient fuel will be consumed during engine start, taxi and run up to confirm the fuel is flowing freely to the engine.

Fuel Tank Selection and Changes

On airplanes requiring fuel tank selection, it is useful to have a set pattern for changing tanks and noting these actions in the fuel log. One method is to change tanks every 30 minutes, but be careful that distractions don't result in you forgetting to change tanks. A reminder mechanism, such as selecting the left tank when the watch hand is between 6 and 12 and the right tank when the hand is between 12 and 6, can be used.

Other considerations regarding fuel tank management are:

> ▸ Do not change fuel tanks just before takeoff.
> ▸ Do not change fuel tanks at low altitudes, such as just prior to landing or on low power descents in case of switching errors (e.g. mistakenly turning to "Off" instead of the "Left" or "Right" tank).
> ▸ Make fuel tank changes when you have sufficient altitude. Before changing a tank while flying, confirm the fuel tank quantity in the tank to be selected and ensure the new valve position is correct by a visual check. Monitor fuel pressure and fuel flow (where gauges are fitted) to ensure there is a satisfactory fuel flow.
> ▸ Double-check the fuel selection lever position before and after changing tanks. Be cognizant of the differences between airplanes, as switching mechanisms can have subtle variations. Human beings are creatures of habit and you may inadvertently place the lever in the wrong position.
> ▸ Do not let a tank run dry. Leaving a quantity of fuel in it ensures you have some fuel left in case there is a malfunction of the selector valve.
> ▸ Plan to make landings with the engines selected to the fullest tank. One technique is to select the fullest tank at the top of descent.
> ▸ Never try to fly on the last cup of fuel.

In-flight Fuel Management

There is a significant fuel penalty if the mixture is not leaned during cruise. There are several reasons, other than fuel consumption, for leaning the mixture, and the procedure outlined in the POH should be followed. In addition, good fuel management requires the following actions:

> ▸ Do not turn the electric fuel pump off too soon after takeoff. Wait until you get to a safe altitude and then monitor fuel pressure/fuel flow after the change has been made.
> ▸ Take positive action if fuel is short (i.e. divert early). Use the fuel log to estimate fuel available.
> ▸ Do not place the airplane in unusual attitudes when the fuel quantity is low (the unusable fuel quantity may increase with abnormal attitudes).
> ▸ Don't be tempted to over fly a previously planned refueling stop (tailwinds have a habit of diminishing just when you need them).
> ▸ Complete pre-landing fuel checks meticulously.

Monitoring Fuel Status

Monitoring the fuel status is best achieved with the aid of a fuel log (see an example below) that enables an on-going assessment of the available fuel margin and fuel reserve. The log should record the quantity of fuel on board at start up and the progressive time available in the tanks. The log can also be used to note fuel tank changes, any re-estimated margins and the impact of variations from the flight plan.

The fuel consumption rate should be monitored during the flight every 30 minutes using the gauges to compare the amount consumed with the estimated usage (you could have a leaking drain valve). Fuel gauge "under readings" should be treated as the actual fuel state, but "over readings" should be ignored. Monitor the time so that you will land within your calculated "time in your tanks".

If your fuel consumption is such that you are likely to use up your reserves before reaching your destination, then land and refuel such that you will definitely have sufficient reserves available for a diversion at your destination. At the end of the flight, check the fuel consumption and compare this with the planned rate so as to achieve more precise planning next time.

FUEL LOG					
Notes		**Fuel Consumption Rate**		38 l/hr	
				Quantity (L/Gal/Kg)	**Time (Minutes)**
		Flight Time		76 litres	120
		Taxi		6.3	10
		Reserves		28.5	45
		Holding		0	0
		Fuel Required		110.8	175
		Usable Fuel		152	240
		Fuel Margin		41.2	65
Time	**Left Tank**	**Right Tank**	**Fuel Remaining**	**Selected Tank**	**Est. Fuel Required**
08:13	120	120	240 min	R	
08:43	120	90	210	L	
09:16	87	90	177	R	
09:46	87	60	147	L	

Avoiding and Handling Weather Hazards

6

Flight into *adverse weather conditions* continues to be a major cause of airplane accidents around the world with many resulting in fatal injuries. Frequently cited reasons for weather related accidents include:

- ▸ Inadequate preflight weather planning and preparation.
- ▸ Not recognizing the hazardous weather warning signs before takeoff.
- ▸ Not recognizing hazardous weather warning signs during flight.
- ▸ Yielding to pressure to get to a destination.
- ▸ Disregarding hazardous weather warnings.

- ▶ Waiting too long before taking avoidance action.
- ▶ Flying into conditions beyond the pilot's capability and experience.
- ▶ A lack of skills to cope with the situation (e.g. lack of instrument flying proficiency).
- ▶ Flying into adverse weather despite ample warning signs.
- ▶ A lack of weather understanding and therefore a lack of appreciation of the risks.
- ▶ An inability to make timely or correct weather decisions.
- ▶ A lack of procedural knowledge to successfully avoid weather hazards.
- ▶ Failing to cope with strong winds, gusting winds or cross winds during takeoff or landing.

This does not mean that a pilot should avoid flying every time there is any likelihood of less than perfect weather. However, it does mean that you must assess and manage the risks. This requires, as a prerequisite, a basic knowledge and understanding of meteorology. In particular:

- ▶ Weather systems and their associated weather characteristics.
- ▶ Localized weather effects and their causes (in particular, terrain effects).
- ▶ The characteristics associated with stable and unstable air.
- ▶ Moisture forming mechanisms.
- ▶ The nature, causes and recognition of hazardous weather.
- ▶ Clouds, and what information they provide about the weather.
- ▶ How to obtain, and understand, all the relevant aviation weather information and data that is available. This should include an appreciation of the basis and limitations of the data.

The objective is not to "outguess" the meteorologist, but to have sufficient background knowledge to understand the available information, and have the ability to assess and manage the weather risks. This involves:

- ▶ Determining what the weather forecast means for the particular flight being planned, taking into account the impact of the weather developing differently from the probable forecast.
- ▶ Making safe judgments and decisions about the flight (e.g. route, altitude, fuel, go/no-go) based on an assessment of the weather as it relates to the capability of the airplane, the capability of the pilot and the options available.
- ▶ Recognizing the weather warning signs and what they mean, both before and during the flight.
- ▶ Taking timely action to avoid hazardous weather conditions.
- ▶ Knowing what action to take should hazardous weather be encountered.

This manual does not attempt to educate the reader on the details of meteorology. There are many excellent texts and publications available that can provide this information. Rather, the objective is to highlight the key issues involved in assessing and managing weather risk.

Avoiding and handling weather hazards requires knowledge, experience, preparation and judgment. As Thomas A. Horne observed in the December 1997 issue of AOPA Pilot, "learning how to deal with the weather must certainly rank as one of the most important skills a pilot can learn".

Making Weather Judgments and Decisions

Many weather-related accidents can be traced back to poor judgments and decisions both before flight and during flight. While the weather information available to pilots nowadays is extensive, pilots must still be able to make educated judgments and decisions about the information and its relationship to the flight. You, the pilot, must make the final decision as to whether to fly or not, and what action to take during the flight.

In making weather judgments and decisions, there are several key points a pilot must remember:

> ▸ The weather is dynamic and can change rapidly.
>
> ▸ Weather forecasting is a complex subject and not an exact science. Forecasts are based on probabilities, with accuracy decreasing the further they extend into the future. Also, some types of weather are more difficult to forecast than others. Predicting the timing of weather systems cannot always be determined with accuracy. Therefore, the weather could well be different from that forecast.
>
> ▸ Published forecasts do not provide all the information available to the forecaster. They are summaries and generalizations of the average conditions and highest probability events.
>
> ▸ Terrain influences weather. While weather forecasts take into account the impact of major features, they don't provide details of the smaller scale modifications caused by local variations in the terrain.
>
> ▸ Regulatory weather minimums are not necessarily safe.
>
> ▸ The degree to which some weather is hazardous depends upon the capability of the pilot and the capability of the airplane. Some weather is hazardous irrespective of the pilot and airplane.

This does not mean that a pilot needs to be a weather expert. However, it does mean that basic meteorological knowledge is required, as well as an ability to make judgments about such things as:

> ▸ What weather is hazardous for a particular flight.
>
> ▸ The likelihood and impact of the weather being different from that forecast.
>
> ▸ Where the best weather is likely to be.
>
> ▸ When to execute escape plans should unacceptable weather be encountered.
>
> ▸ What information is needed in order to make safe judgments and decisions.
>
> ▸ The possibility and nature of any localized weather phenomena.

Incorrect judgments have been made on many occasions when pilots have attempted, or continued flight, into conditions beyond either their capability, or that of the airplane. Fortunately, weather judgment can be learnt through training, education and experience. Experience is particularly important. Analysis of accident statistics by the AOPA Air Safety Foundation (*Safety Review – General Aviation Weather Accidents*) indicates that the more experience you have, the lower the risk of having a weather related accident. However, it does not make you immune.

Improving Your Weather Judgment and Decision Making Capability

Weather education, training and experience is not a once off exercise. It is a never-ending journey that is helped by:

> ▸ Being aware of your own limitations and guarding against over-confidence.
>
> ▸ Acquiring and maintaining a good understanding of aviation meteorology, weather briefings and hazardous weather recognition, by attending seminars and pursuing self-study.
>
> ▸ Keeping up to date with the weather information sources, what information they provide and their limitations.
>
> ▸ Having knowledge of weather avoidance strategies and tactics.
>
> ▸ Developing better weather sense by regularly assessing the overall weather picture through analysis of available weather reports and monitoring of the actual weather.
>
> ▸ Practicing weather decision making by periodically planning a hypothetical flight and going through all the required motions including weather briefing, flight planning and then seeing how the forecast develops.
>
> ▸ Practicing landing and takeoff with an instructor in weather conditions not normally encountered (e.g. gusting wind, cross wind).
>
> ▸ Flying in marginal VFR weather conditions with an experienced, and appropriately qualified, instructor to gain exposure to more demanding weather and to practice making weather decisions.
>
> ▸ Flying with a more experienced pilot in weather conditions you have not previously encountered.
>
> ▸ Reviewing weather accident reports and ascertaining how the accident could have been avoided.

Every pilot's judgment and decision making capability will be different, depending on their knowledge and experience. The key is to continue to learn, and to fly within your limitations by setting personal weather minimums that are commensurate with your knowledge and experience.

Preflight Weather Planning and Preparation

The importance of weather planning and preparation should never be underestimated. Many weather-related accidents have had their origin in inadequate preflight planning and preparation. In some instances the weather briefing was either not obtained, incomplete or totally disregarded. Weather planning must, therefore, be considered to be as important as any other preflight activity.

Extent of the Preflight Weather Briefing

How elaborate a preflight weather briefing should be, depends upon the nature of the flight and the prevailing weather. However, weather forecasts are rarely 100% accurate and can change rapidly. Even for a relatively short flight, a thorough briefing is generally warranted. In any case, if you are less experienced, it is good practice and educationally beneficial.

Obtaining Weather Information

It is best to begin by developing an awareness of the overall weather picture two to three days before the planned flight. Information can be obtained from TV, radio, newspapers and the Internet (e.g. satellite images, synoptic charts, radar images). Note should be taken of the following, in order to develop a mental picture of the likely weather along the intended route:

> - The systems that are influencing the weather.
> - The location of fronts and how fast they are moving.
> - The expected movement of the pressure systems.
> - The possibility of hazardous weather.
> - The direction of worsening weather.
> - The direction of improving weather.
> - The stability of the atmosphere.
> - The overall wind flows.
> - The moisture content in the atmosphere.
> - The overall weather trend.

For longer flights it is advisable to get an outlook weather briefing on the day before the flight to ascertain how the weather is developing.

On the day of the flight it will be necessary to obtain both the actual, and forecast weather conditions for the departure point, en-route, destination and alternates (e.g. Area Forecasts, Terminal Forecasts, Pilot Reports). Obtaining actual weather gives you an indication of how the weather is progressing relative to the forecasts. It is also good practice to obtain the weather information in regions adjacent to the route being flown to compare forecasts and identify what weather may be coming (a marked change may be approaching). You may need to speak to a meteorologist depending on the circumstances. Make sure you use all the relevant sources of weather information such as weather reports, synoptic charts and satellite images, to get a total picture of the weather, rather than just one perspective. The more doubtful the weather, the more information you need in order to be able to make the best possible assessment.

Assessing and Interpreting the Weather Information

The gathered weather information needs to be interpreted into information that is relevant and pertinent to the flight being planned. Asking the following questions can help this process:

- What is the current weather and what is predicted to happen?
- Is there any frontal activity and how will movement of the front affect the weather?
- What are the conditions before and after a front (it may move faster or slower)?
- Is the weather weakening or intensifying?
- What is the wind doing at various levels?
- Where is the freezing level relative to the cloud base?
- What is the visibility?
- What is the consequence of the weather moving faster or slower than forecast?
- Is there any unstable weather along or adjacent to the intended route?
- What hazardous weather is forecast along the route?
- Is any thunderstorm activity, potentially affecting the flight (frontal, embedded, isolated)?
- What is the impact of the terrain over which the flight is being planned?
- What warning signs would indicate that the weather is getting worse than forecast?
- How might the weather develop differently from that forecast?

Obtaining answers to these questions will assist the pilot in making the required preflight weather related judgments and decisions.

Making the Required Judgments and Decisions

The next step in the process is to ask yourself a number of additional questions that will help in the decision making process:

- Is the expected weather beyond my capabilities , taking into consideration visibility, turbulence, cloud and other potential weather hazards?
- Are my personal standards or minimums consistent with my experience, proficiency and knowledge?
- Is the airplane suitably equipped to handle the likely weather conditions?
- Is the terrain a factor?
- What are my options in case the weather deteriorates?
- What route should I fly to maximize safety?
- What cruise levels will provide the safest flight?
- What fuel do I need to carry to account for various possible weather eventualities?
- Have I obtained all the necessary information?
- Should I delay the flight or possibly not go at all?

In asking these questions, the following weather risk reduction strategies should be taken into consideration:

> ▸ Avoid areas where the trend is for weather deterioration to unacceptable conditions.
> ▸ Do not try to beat the weather.
> ▸ If possible, fly in a direction of improving weather.
> ▸ Be cautious about fuel calculations and use the strategies outlined in Section 5.
> ▸ Be prepared in your mind that a cancellation/diversion/return/landing may be necessary. Eliminate any notion of "get-thereitis". Successful flying demands your acceptance that you may not always reach your destination, may have to divert or may not be able to fly on the day. Make sure your passengers understand and accept these possibilities.
> ▸ If you are not proficient at instrument flying, factor this into your options.
> ▸ Be especially cautious at night, as the options are fewer. Avoid night flying when there is thunderstorm activity.
> ▸ Plan your trip early in the day (thunderstorm activity generally occurs in the afternoon and evening but note that this is not always the case).
> ▸ Plan to fly visual if there is a chance of thunderstorms. Alternatively, plan a different route or cancel the flight.
> ▸ When flying IFR, if the weather is below the departure airport's approach minimums, don't takeoff (as you may not be able to land in the event of a problem just after takeoff).
> ▸ Be cognizant of the effect of terrain and plan a route that avoids terrain-induced hazards.
> ▸ Be cautious about flying VFR on top of clouds unless the clouds are widely scattered and are expected to stay that way.
> ▸ Make certain you have pre-planned escape routes in case of deteriorating weather.

Then, on the basis of your assessment of the information available, make the necessary judgments and decisions as to:

> ▸ Whether the flight should be delayed or cancelled.
> ▸ Whether the route and/or altitude should be changed.
> ▸ Whether an intermediate stop will be necessary or is likely.
> ▸ What escape routes are possible should the weather deteriorate.
> ▸ Whether the conditions exceed your personal minima.
> ▸ Whether to fly IFR or VFR (if IFR qualified).

In making these weather judgments and decisions, it is worth noting the point made in Issue 6/83 of Transport Canada's Aviation Safety Letter, in which readers were reminded to "make the weather decision while you still have a choice".

Hazardous Weather Warning Signs

The ability to recognize and understand hazardous weather warning signs, both before and during flight, provides the foundation for making timely and safe weather-related decisions. Weather impacts on visibility, airplane controllability, aerodynamic performance, structural integrity and engine performance. Weather becomes hazardous when limits associated with these pilot, and airplane, parameters are exceeded. Some weather is hazardous irrespective of the circumstances.

The following weather conditions can be potentially hazardous to flight:

- Turbulence (gusting winds, rough air, eddies, updrafts, downdrafts, rotors).
- Cross winds.
- Wind shear.
- Thunderstorms.
- Microbursts.
- Cloud.
- Fog.
- Rain and drizzle.
- Wind (high wind, gusting wind).
- Freezing rain.
- Hail.
- Lightning.
- Haze.
- Smoke.
- Sun glare.
- Snow (flakes, pellets, ice crystals, ice pellets).
- Frost.
- Dust.
- Below freezing temperatures and moisture (airframe icing).
- Humidity (carburetor icing).

The weather conditions that may be encountered are a function of the weather system(s) and any influence of the earth's topography. The nature of the earth's surface and its interaction with the atmosphere, influences the weather both on a macro and micro scale. Fortunately, there are numerous signs that provide warning of weather conditions that may be hazardous to your flight. These can be categorized as:

- Warning signs inherent in the weather forecast.
- Warning signs that relate to the nature of the terrain.
- Warning signs that are observed from the cockpit.
- Other warning signs.

Awareness of these signs is essential for safe flying.

Forecast Warning Signs and Hazards

While forecast data and information is nowadays extensive, it must be interpreted and understood in the context of the flight being planned. An appreciation of the weather systems that signal potentially hazardous weather conditions will help in guarding against an inappropriate weather assessment. Some examples of the more common weather situations are given below.

Forecast Warning Sign (weather systems)	Possible Hazardous Weather
Cold Fronts Cold fronts are associated with many kinds weather, move relatively fast and are characterized by cumuloform clouds. They have a comparatively narrow band of weather and may have rapid changes in wind, temperature and humidity. Cloud can form well ahead of a front but clearer conditions generally exist after frontal passage. Precise timing of a front is difficult to predict.	▸ Gusty turbulent winds with a wind shift after frontal passage. ▸ Low cloud. ▸ Wind shear, which can exist for some time after frontal passage. ▸ Precipitation, heavy at times. There can be a significant rain band ahead of the front during winter. ▸ Possible thunderstorms in a fast moving front. ▸ Possible radiation fog after frontal rain.
Warm Front Warm fronts involve a wider band of weather than cold fronts and are characterized by stratiform clouds.	▸ Generally widespread precipitation. ▸ Wind change with frontal passage. ▸ Significant amount of cloud both before and after frontal passage.
Winter Warm Fronts Adapted from diagram courtesy of CAA of New Zealand	▸ Icing above the freezing level. ▸ Freezing rain/drizzle. ▸ Snow. ▸ Ice pellets.

Forecast Warning Sign (weather systems)	Possible Hazardous Weather
Stationary Front Slow moving or stationary fronts provide time for moisture to be gathered.	▸ Low ceilings. ▸ Poor visibility. ▸ Widespread light precipitation. ▸ Freezing rain or drizzle during winter.
Occluded Front Occluded fronts generally contain weather associated with both cold and warm fronts.	▸ Steady, light to moderate precipitation. ▸ Low cloud. ▸ Possible embedded thunderstorms.
Upper Level Trough Troughs can create and intensify disturbances. They provide a lifting mechanism for unstable air below, generally resulting in poor weather.	▸ Cloud. ▸ Precipitation. ▸ Turbulence in upper layers.
Temperature Inversion Temperature inversions occur when warmer air overlies cooler air due to radiation or frontal action. 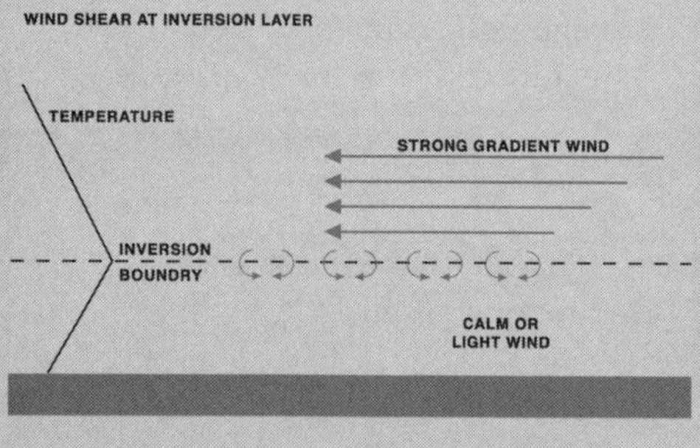	▸ Unstable air. ▸ Wind shear at the inversion line. ▸ Stratus cloud and smog trapped below the inversion line.
Temperature/Dew Point Spread A close spread indicates a high likelihood of fog and haze.	▸ Fog. ▸ Haze.

Forecast Warning Sign (weather systems)	Possible Hazardous Weather
Thunderstorms The effects of thunderstorms can extend up to 20 nautical miles from the center.	▸ Extreme turbulence in and beneath the cloud. ▸ Wind shear many miles from the storm. ▸ Heavy rain. ▸ Lightning. ▸ Hail. ▸ Microburst. ▸ Severe updrafts/downdrafts. ▸ Low ceilings. ▸ Icing.
Tropical Cyclones/Hurricanes Tropical cyclones reach their greatest intensity over warm tropical waters but weaken as they move inland. NOAA National Climate Data Centre	▸ Extreme hazardous conditions.
Tornadoes Usually associated with severe thunderstorm activity.	▸ Very high rotational winds and extreme hazardous conditions.

Terrain Induced Weather Warning Signs and Hazards

Weather that is influenced by terrain features includes:

- ▶ Differential thermal effects such as sea breezes and land breezes.
- ▶ Mountain and valley winds.
- ▶ Mountain waves and rotors.
- ▶ Eddies, wind-shifts and wind shadows caused by trees, buildings etc.
- ▶ Terrain contours effects resulting in funneling and updrafts or downdrafts.

The following are some typical examples:

Terrain Conditions and the Possible Induced Weather Hazards	Warning Signs
Mountains or high ground with the wind at approximately 90° to the ridge line ▶ Mountain waves. ▶ Rotors or eddies. ▶ Lee wave turbulence, updrafts and downdrafts. Lenticular and rotor clouds may be present, but not always. The effect can occur over relatively low ridges and can extend for a considerable distance downwind, depending on the wind strength. Note that local factors can affect the intensity of rotors and eddies.	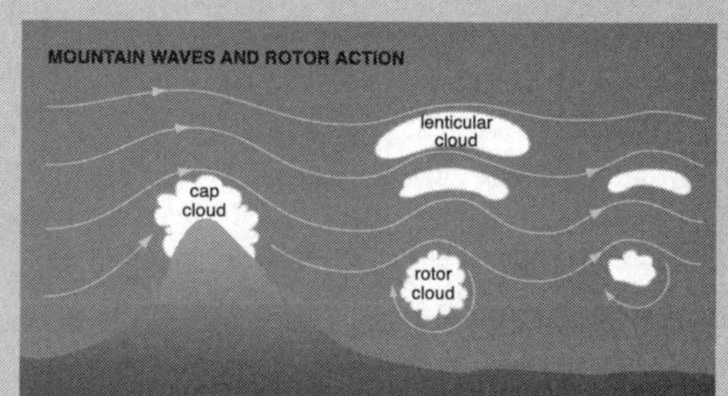Adapted from diagram courtesy of CAA of New Zealand
Trees and buildings near runways ▶ Wind shadows. ▶ Wind shear. ▶ Eddies. ▶ Changes in wind direction.	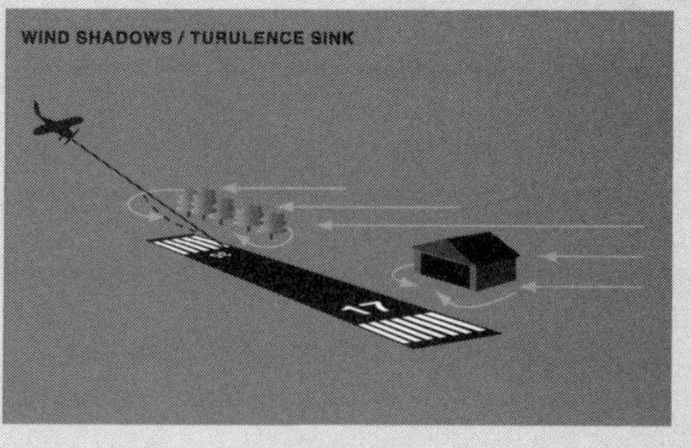

Terrain Conditions and the Possible Induced Weather Hazards	Warning Signs

Terrain contour effects

- ▸ Wind funneling in valleys (up valley during the day and down at night).
- ▸ Wind concentration and funneling between hills and mountains.
- ▸ Updrafts and downdrafts caused by sloping terrain.
- ▸ Small-scale turbulence in the lee of a hill.

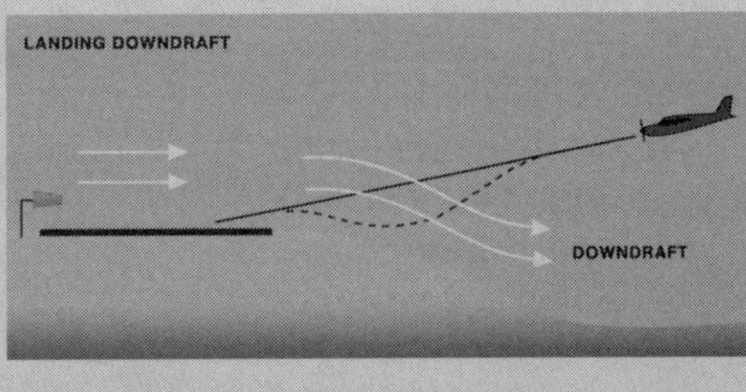

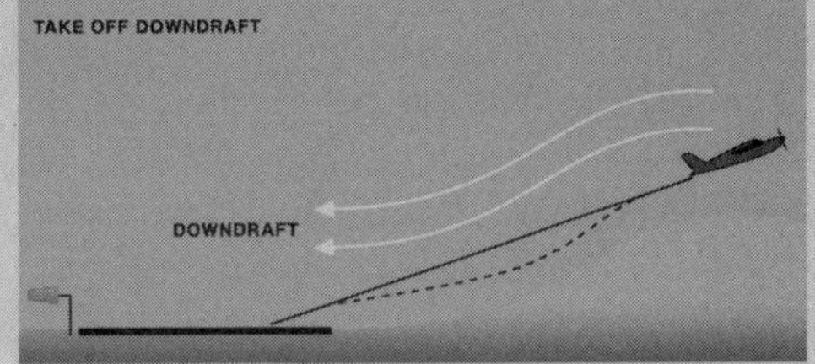

Terrain/cloud effect - Lampshade effect

A lampshade effect occurs when low light under an overcast sky reduces visibility to such an extent that the pilot can only see objects clearly within a narrow, restricted cone.

Adapted from diagram courtesy of Flight Safety Australia

Terrain/cloud effect - Low cloud and rising terrain

As the result of poor light under thick cloud, a pilot places increasing emphasis on ground features and does not notice an increase in terrain height.

Adapted from diagram courtesy of Flight Safety Australia

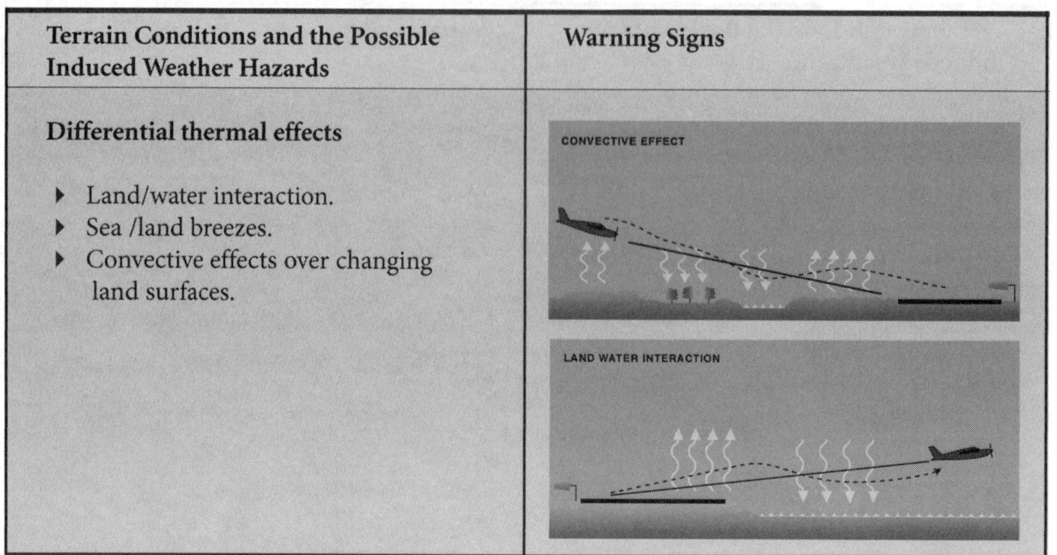

In-flight Observed Weather Warning Signs and Hazards

By observing various weather indicators, a pilot can be warned of possible hazardous weather conditions during flight. The cloud type, shape, color and density are particularly important indicators, in that they tell a lot about the stability of the atmosphere, the level of moisture, the degree of turbulence and flight altitude limits. Other indicators include changes in wind direction, wind speed, visibility, temperature, distinctiveness of the horizon and the presence of rain.

For example, towering cumulus clouds can indicate the early stages of a thunderstorm while the presence of virga can indicate the possibility of severe wind shear. Cloud and visibility are in fact very useful indicators of deteriorating weather conditions. Changes in wind direction combined with increasing cloud, may indicate the presence of a front and its associated weather. An increase in wind speed may indicate the approach of a low pressure system and its typically associated poorer weather.

Such indicators provide valuable information to the pilot about potentially deteriorating weather conditions, thus permitting timely decisions to be made as to what avoidance measures should be taken. Some examples of typical indicators follow.

Warning Sign and the Possible Weather Hazard(s)	Typical In-flight Observation
Gradual lowering of the cloud base or rising terrain ▸ Reduced terrain clearance. ▸ Possible widespread precipitation. ▸ Reduced visibility. ▸ Lack of a distinct horizon.	 Courtesy of Jeremy Zawodny
A line of heavy dark clouds or darkening clouds. Hazards depend on the type of cloud. If it is convective weather, then possible hazardous weather can include the following: ▸ Severe turbulence. ▸ Poor visibility. ▸ Hazardous landing conditions. ▸ Precipitation. ▸ Hail.	 Jimmy Deguara
Towering cumulus ▸ Turbulence (moderate to severe). ▸ Possible precipitation. ▸ Early stage of a thunderstorm.	2008 © University Corporation for Atmospheric Research

Warning Sign and the Possible Weather Hazard(s)	Typical In-flight Observation
Opening in a wall of dark cloud ▸ Dangerous turbulence. ▸ Precipitation. ▸ Poor visibility.	 Courtesy of Commonwealth Bureau of Meteorology
Lenticular clouds ▸ Strong mountain wave action. ▸ Rapidly descending air on lee side of ridge. ▸ Severe turbulence below cloud.	 NOAA Photo Library/US National Parks Service
Cumulonimbus clouds ▸ Severe to extreme turbulence. ▸ Wind shear many miles from the storm. ▸ Violent turbulence in and beneath the cloud. ▸ Heavy rain. ▸ Hail. ▸ Microbursts.	 Courtesy of Commonwealth Bureau of Meteorology

Warning Sign and the Possible Weather Hazard(s)	Typical In-flight Observation
Mammatus clouds ▸ Extremely high water or ice content. ▸ Hail or intense rain. ▸ Severe turbulence below the cloud base.	 NOAA Photo Library/NSSL
Cumulus clouds ▸ Possible wind shear at low levels (gusty surface winds). ▸ Light to moderate turbulence. ▸ Rough ride. Hard angular contours indicate more turbulent conditions.	 Courtesy of Commonwealth Bureau of Meteorology
Virga ▸ Indication of a possible dry microburst. ▸ Severe wind shear.	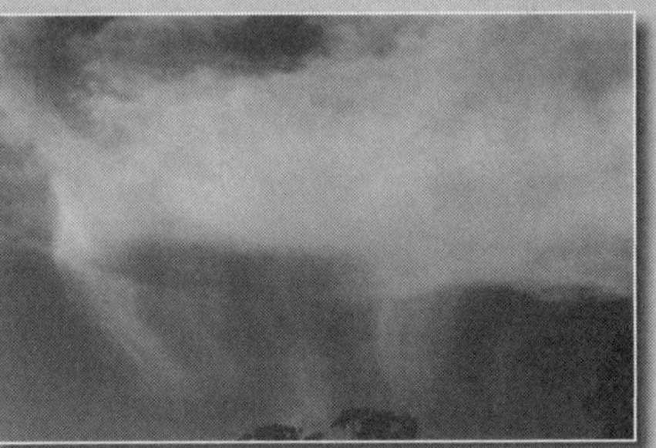 Michael Bath

Warning Sign and the Possible Weather Hazard(s)	**Typical In-flight Observation**
Rain ▸ Reduced visibility. ▸ Possible deteriorating conditions. ▸ Slippery runways.	 Courtesy of Commonwealth Bureau of Meteorology
Lightning ▸ Thunderstorms. ▸ Possible electric shock if the airplane is not well bonded. ▸ Possible airplane damage. ▸ Pilot disorientation and possible temporary blindness. ▸ Instrument failure.	 NOAA Photo Library/NSSL
Fog ▸ Reduced visibility . For example, the airfield may be visible while overflying, but may not be visible on final approach.	 Michael Bath

Warning Sign and the Possible Weather Hazard(s)	Typical In-flight Observation
Blowing dust ▸ Turbulence. ▸ Poor visibility.	Katsuhiro Abe, Japan Meteorlogical Agency (Courtesy of Commonwealth Bureau of Meteorology
Snow ▸ Reduced visibility. Snow can sometimes reduce visibility very quickly. ▸ Indistinct horizon and whiteout. ▸ Unrecognizable landmarks. ▸ Hazardous takeoff and landing conditions.	NOAA Central Library

Less Obvious In-flight Weather Warning Signs

Possible Hazard(s)	Warning Signs
Wind shear Wind shear is a form of turbulence where there is abrupt loss or increase in wind speed. This results in fluctuating airplane airspeed and deviations from the flight path. ▸ Possible undershoot. ▸ Possible overshoot. ▸ Possible stall.	▸ Rising dust or sand. ▸ Obstructions to wind flow. ▸ Divergent windsocks or smoke. ▸ Strong shafts of rain. ▸ Virga. ▸ Divergent wind patterns indicated by grass, trees etc.

Possible Hazard(s)	Warning Signs
Strong/gusty surface winds ▸ Low level turbulence. ▸ Wind shear.	▸ Horizontal/moving windsock.
Strong upper winds with little or no wind at ground level ▸ Wind shear. ▸ Turbulence.	▸ Drift correction at altitude with little or no windsock indication.
Change in wind direction or speed ▸ Higher fuel consumption. ▸ Drift off course.	▸ Change in drift correction. ▸ Change in smoke or dust movement. ▸ Change in ground speed. ▸ Change in the prevailing meteorological conditions. ▸ Transition to an area of lower pressure with associated poorer weather conditions.

Other Warning Signs

Warning Sign and the Possible Weather Hazard(s)	Observation
Summer ▸ Potential for afternoon and evening thunderstorms.	NOAA Photo Library/NSSL
Early morning ▸ Possible fog. ▸ Poor visibility (flying into a rising sun).	2008 © University Corporation for Atmospheric Research
Late in the day ▸ Marginal weather can get worse. ▸ Poor visibility (flying into a setting sun).	Michael Bath

Avoiding Hazardous Weather

Once a decision has been made to depart, irrespective of whether you are flying VFR or IFR, you need to continually re-assess the weather and take avoidance action as necessary. Information about the weather is obtained from observations both inside and outside the cockpit, together with weather updates from the available flight services. Any action you take will depend on your assessment of the information and your knowledge of hazardous weather avoidance strategies and tactics. The assessment process can be helped by regularly asking yourself a series of questions during the flight:

> ▸ What is the weather doing? Is it deteriorating? Are there any hazardous weather warning signs?
> ▸ Have I obtained updates of the weather en-route and at the destination?
> ▸ What are the options if the weather deteriorates, taking into consideration such factors as fuel endurance, terrain, landing sites and last light? Have I left myself enough options?
> ▸ Am I likely to exceed my personal minimums?
> ▸ Am I likely to exceed the capability of the airplane?
> ▸ Can I continue the flight as planned with safety?

Make sure you are familiar with the indicators that can be used to identify when weather conditions are deteriorating during flight. Expect the weather to change (irrespective of the forecast), so that you will always be ready to take avoidance action if necessary.

General Hazardous Weather Avoidance

Weather that needs to be avoided depends on the capabilities of the pilot and the airplane. However, irrespective of these capabilities, there are a number of general tactics that can be applied to minimize the risk of encountering hazardous weather.

> ▸ Spend sufficient time looking up at the weather, searching for hazardous weather warning signs. In addition, keep an eye on the weather behind you.
> ▸ If flying VFR:
> > ▸ Watch for a steadily lowering cloud base to avoid being forced to descend to a dangerously low altitude to remain visual. Don't assume it is only temporary.
> > ▸ Watch for rising terrain (this effectively lowers the cloud base).
> > ▸ Watch for a steadily rising cloud mass (when flying on top) that could result in the airplane being engulfed by building cumulonimbus cloud. The situation is unlikely to arise if you only fly on top when the clouds are widely scattered.
> > ▸ Stay sufficiently below the cloud base to maintain a clear horizon.
> ▸ Monitor the ground speed and drift, to detect changes in wind direction and strength.
> ▸ Maintain a continuous watch on your possible pre-planned escape routes and have a willingness to use them if necessary. Never place yourself in a position where your only option is to continue into deteriorating weather.

The key tactic is to monitor the weather, and before it gets too bad (i.e. hazardous to your flight), take action to change altitude, detour, return, hold, divert or land. Make the decision early. If necessary, request ATC to identify an area of better weather or seek navigational guidance. Don't be lulled into a false sense of security by any earlier favorable forecasts.

Avoiding Flying VFR-into-IMC

Courtesy of Australian Transport Safety Bureau

"Two of the factors that may have contributed to this accident, were that the weather along the intended route was not suitable for visual flight and the pilot attempted to conduct the flight visually below cloud"

While VFR pilots may have some exposure to instrument flying, it is wise not to rely on that limited ability to get you out of trouble should IMC be entered inadvertently. The best strategy is to avoid entering cloud by:

- Never beginning a flight on the basis of "hoping that the weather will improve".
- Avoiding hazardous attitudes such as "it won't happen to me" and "get-thereitis".
- Having a good appreciation of what weather is hazardous and, if unsure, seeking appropriate advice.
- Having sufficient knowledge and experience to be able to recognize any hazardous weather warning signs.
- Setting go/no-go decision criteria consistent with flying knowledge and experience.
- Always looking for the earliest signs of deteriorating weather and being mentally prepared to divert or turn back, without delay.

The right attitude and familiarity with the weather, combined with good preflight planning and in-flight judgment, will significantly reduce the risk of a VFR-into-IMC accident.

Avoiding Thunderstorms in Flight

If there is a probability of thunderstorms, then one or more of the following well known avoidance tactics should be adopted during flight:

- Vivid and frequent lightning indicates the probability of severe thunderstorms. Therefore, divert to avoid.
- Stay visual if possible, unless you have the appropriate weather detecting aids and understand their use and limitations. Note that lightning detectors and weather radar are aids to help you avoid, or get around thunderstorms, not through them. Even then, they have their limitations.
- Try to avoid thunderstorms by at least 20 nautical miles.
- Try to stay upwind of a thunderstorm and use movement of the weather to your advantage.
- Get a radar update if possible.
- If you must divert downwind, put one nautical mile for every one knot of wind, at your altitude, between you and the storm.
- Some "Don'ts":
    - Don't land or take off in the face of an approaching thunderstorm.
    - Don't attempt to fly under a thunderstorm, even if you can see through to the other side.
    - Don't fly into a cloud mass containing possible embedded thunderstorms without airborne radar (but note that weather radar only detects precipitation drops, not turbulence).
    - Don't fly on if thunderstorms bar your way.
    - Don't trust the visual appearance of a thunderstorm as a reliable indicator of turbulence inside a thunderstorm.
    - Don't try to circumnavigate frontal thunderstorms, as they cover considerable ground.
    - Don't assume that a thunderstorm will always move in the direction of the prevailing wind.
    - Don't try to slip through a hole in a line of thunderstorms, as it can quickly close.
    - Don't attempt to fly on top of cumulus clouds when cumulonimbus clouds are nearby. A cumulus cloud can grow rapidly, and at a rate that may exceed the climb performance capability of the airplane.
    - Don't try to race a thunderstorm to the airport.

Thunderstorms are highly dangerous and avoidance is the best strategy.

Avoiding Mountain (High Ground) Turbulence

Mountain, or high ground, induced turbulence can be severe. However, there are tactics that can be adopted to significantly reduce the risk.

> - Select a cruising altitude above the highest ridgeline, at least equal to the height of the ridge above the terrain.
> - If possible, fly to the windward side of ridgelines and mountain ranges.
> - Avoid flying in the vicinity of rotor clouds.
> - Try to approach the mountain or high ground at an oblique angle (45°) with a predetermined escape path. This offers:
>   - The best view of the approach.
>   - A shallow approach angle, and therefore a shallow escape angle (minimum angle of bank required).
> - If the wind at the ridge is greater than 20 knots, avoid flying downwind of high ground if you cannot climb to a high enough altitude.

Flying in Hazardous Weather Conditions

There will be many occasions during your flying career when you will encounter weather that can be potentially hazardous to your flight. Much of this, such as gusting wind, turbulence and crosswind, can be readily managed, provided they do not exceed your capability or that of the airplane. However, at some point, more severe conditions could be encountered, and being well prepared (i.e. knowing what to do) will enable the situation to be handled with safety.

Managing Hazardous Weather During Takeoff and Landing

A significant number of weather related accidents have occurred during takeoff and landing. Therefore, before takeoff or landing make sure you have a good understanding of the weather conditions, and determine whether they pose a risk or not relative to your proficiency and experience. In addition, be familiar with the general takeoff and landing risk reduction tactics outlined in Section 8, and be cognizant of the need to:

> - Compensate adequately for the wind (e.g. ailerons into wind, sufficient rudder to maintain directional control, higher airspeed in gusting conditions).
> - Maintain sufficient airspeed above the stall speed.
> - Maintain proper descent rates.
> - Be prepared for wind shear.

If there is uncertainty about the conditions, or the approach is not going to plan, apply good judgment and go around, divert or carry out a missed approach.

Managing Low Level Wind Shear and Turbulence During Descent and Approach

Low level wind shear and turbulence is always a possibility during descent and approach when conditions such as strong temperature inversions, frontal activity, thunderstorms, sea breezes, microburst and strong or gusty surface winds, exist. Using the following tactics can reduce the risks associated with such turbulence:

> ▸ Slow the airplane down to maneuvering speed.
> ▸ Fly smoothly.
> ▸ Keep descent rate to 500 feet per minute or less.
> ▸ Do not make rapid altitude or heading changes.
> ▸ Increase the approach speed by one half of the Gust Factor if the wind is gusting or wind shear is forecast. Limit increase to 10 knots, otherwise a long float and control problems can result. For example:
>> Surface wind 15 knots gusting to 25 knots
>> Gust Factor = (25-15) = 10 knots
>> Therefore, increase approach speed by 5 knots.
> ▸ Use only partial flaps so as to improve the stability of the approach.
> ▸ Be prepared to go around if necessary.
> ▸ Watch out for wind shear, and if detected, immediately add power, achieve climb speed and consider diverting.

If You are Flying VFR and Enter IMC

Flying VFR into IMC can be avoided. However, if an inadvertent situation arises, the following well-recognized tactics should be followed:

> ▸ Relax and take stock of the situation. In particular, note your height above ground level.
> ▸ Transfer your attention to the instruments , level the wings and note the heading.
> ▸ Turn on and use the autopilot for straight and level flight.
> ▸ There are several options depending on the weather situation and your level of skill/proficiency at instrument flying.
>> ▸ If VMC is ahead, fly straight and level on instruments.
>> ▸ If the cloud is not too thick, climb at a constant heading.
>> Alternatively descend, if you are well above the lowest safe altitude.
>> ▸ If the cloud is too thick, or it is a better option, carry out a level rate one 180° turn.
> ▸ Communicate your predicament with ATC.
> ▸ Make small control inputs and keep angle of bank to rate one or less.
> ▸ Turn on the pitot heat.
> ▸ Keep your head still.

If Caught Near or In a Thunderstorm

The best strategy is to avoid a thunderstorm. However, if you do get caught, here are the recommended actions to take:

> - Stay relaxed and calm and be prepared for extreme turbulence, rain, hail, snow, icing, lightning and wind shear.
> - Prepare any passengers for the likely conditions.
> - Tighten seat belts, shoulder harness and secure all loose objects.
> - Turn on the pitot heat and carburetor heat (plus any other anti-icing equipment).
> - Fly at an altitude below the freezing level and ensure adequate terrain clearance.
> - Turn up the cockpit and instrument lights.
> - Slow to maneuvering speed.
> - Set the power to give the recommended speed and avoid changing power settings unless absolutely necessary.
> - Use smooth, moderate control inputs.
> - Don't chase altitude or airspeed. Try to maintain a constant attitude using the attitude indicator and let the airplane ride through the turbulence. Control the attitude regardless of anything else and disregard conflicting indications on other instruments.
> - Keep wings level and maintain heading (turning imposes additional airplane stresses).
> - Fly the most direct route through the storm.
> - Focus on flying using the instruments, and don't look outside (avoids temporary blindness from lightning).
> - Don't turn back once the thunderstorm has been entered.

Dealing with Fog

Thin layers of fog can produce an illusion of adequate visibility when flying over the field, but significantly impaired visibility when on the landing approach.

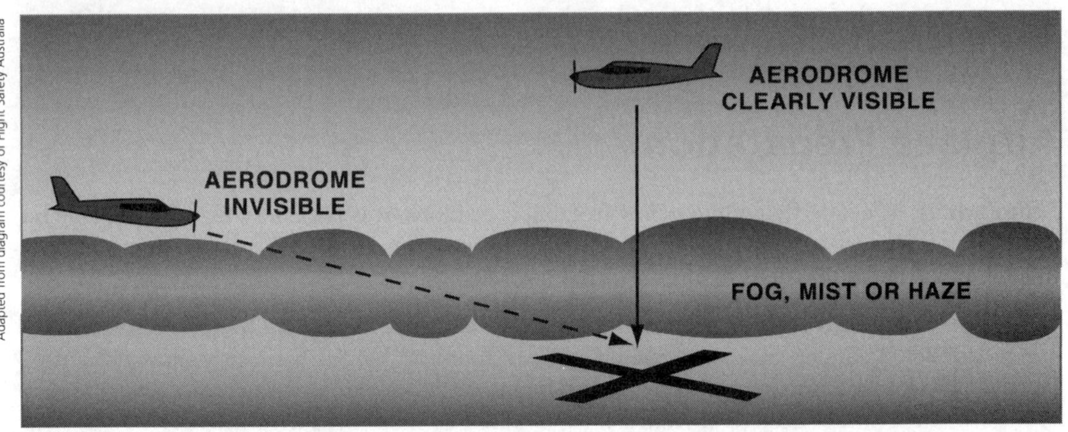

Pilots need to be aware of this possibility and execute a go around immediately for a diversion or hold. Some advice on dealing with fog is as follows:

> ▸ If fog is probable, make sure you arrive well before dark in case of ground fog that has not been forecast.
>
> ▸ If fog has been forecast make sure you have an alternate or wait until it has burned off completely.

Winter Flying in Cold Climates

Flying in conditions where the temperature regularly falls below freezing poses additional risks that need to be understood and managed. Airframe icing, carburetor icing, icy runways, slush, gusty winds, blowing snow, drifting snow, poor visibility, whiteout and changeable weather are just some of the hazards you may encounter. Pilots therefore require:

> ▸ An airplane that has been suitably prepared for winter operation.
>
> ▸ Intimate knowledge of the airplane's winter operating procedures and systems.
>
> ▸ Proficiency at handling the likely conditions, both in the air and on the ground.
>
> ▸ Cold weather knowledge including the skills to be able to recognize deteriorating weather warning signs.
>
> ▸ An ability to use extra caution and exercise good judgment.

This section highlights some of the key issues. If you are unfamiliar with winter flying, it is prudent to seek advice from operators and airplane engineers who are permanently located in colder climates and have relevant knowledge and experience.

Allow Adequate Time

Because of the low temperatures in winter, there is a tendency to rush preflight activities. However, the very opposite is needed. Experience has shown that extra time is required for everything. A good rule of thumb is to allow twice as much time as you do in summer.

Airplane Preparation

Before winter, or before flying into below freezing conditions, it is necessary that the airplane be appropriately winterized. This must be done strictly in accordance with the manufacturer's recommendations. Factors that need to be considered include:

> ▸ Using the correct grade of oil for the expected temperature conditions.
>
> ▸ Some engines may need manufacturer's approved winter baffles to allow oil and cylinder temperatures to reach and maintain the correct operating levels. If fitted, the cylinder head temperature must be carefully monitored.
>
> ▸ Checking the condition of all hoses, flexible tubing and seals for deterioration.

- ▶ Ensuring that all oil lines and tanks are appropriately insulated to preclude the possibility of oil congealing.
- ▶ Adjusting the control cables to account for the colder conditions.
- ▶ Checking the condition of the heater and defrost system for proper functioning and leak tightness. Use a carbon monoxide detector as extra insurance.
- ▶ Ensuring the battery is in good condition, fully charged and removed from the airplane if parked outside for a lengthy period.
- ▶ Removal of the wheel covers to reduce the chance of frozen slush locking up the wheels and brakes.
- ▶ Checking the operability of the anti-ice and de-ice systems.
- ▶ Checking engine idle speed and carburetor heat operability.
- ▶ Ensuring no drain holes are obstructed.
- ▶ Checking the crankcase oil breather lines are clear and free of freezable material.
- ▶ Checking the requirements for any special lubricants.

If the airplane is parked outside, consideration should be given to the use of covers for at least the pitot tube, engine and wings. Also, in winter there is a greater risk of water condensing in the fuel tanks, and it is good practice to leave the tanks full while parked.

Preflight Planning

Apart from ensuring that the airplane is properly prepared, a number of preflight planning activities need particular attention:

- ▶ In winter, the number of daylight hours is limited. Leave adequate time.
- ▶ Planning should include a check on your winter flying knowledge, the recommended safe operating procedures and your proficiency at handling the likely conditions.
- ▶ Icing is a major risk during winter, particularly from freezing rain. Careful planning of the route and altitudes is required, along with knowledge of how to deal with icing if inadvertently encountered. If the airplane is fitted with anti-ice and de-ice equipment, then familiarity with its operation is needed. However, don't be misled into thinking it is effective in all icing conditions.
- ▶ Careful consideration and planning of escape routes is required in case of icing problems or unexpected weather.
- ▶ Weather in winter can change rapidly and is very unforgiving. Careful weather planning and assessment is crucial. This includes carrying adequate fuel reserves, as weather can vary significantly from that forecast, and runway conditions can easily force a diversion. Alternates should be pre-planned.
- ▶ Check that the destination, intermediate and alternate aerodromes are in fact open.
- ▶ NOTAM's should be checked for advice on winter maintenance, which may not be readily visible from the air.

As mentioned in an earlier section, attitude is a potential killer. File a flight plan, set reasonable limits and avoid "get-thereitis". Think very carefully about the go/no-go decision.

Personal and Passenger Preparation

Both the pilot and passengers should either dress or, have clothing available, for the worst weather along the intended route. Clothing should include good footwear and head protection, but a check should be made on the ability to handle the airplane properly in heavy winter clothing (if being worn).

Personal equipment should include sunglasses and survival gear appropriate to the individual needs, temperature and route being flown.

Preflight Inspection and Checks

Winter flying requires a very thorough preflight inspection and some additional checks.

- When refueling, make sure the airplane is properly earthed. The very low humidity on a crisp, cold day can be conducive to a build up of static electricity.
- When taking fuel samples, check for water and contamination and ensure adequate quantities are taken from each drain point. If it doesn't drain freely, suspect ice or sediment in the line or sump. It may be necessary to move the airplane into a heated hanger to melt any ice.
- Check for accumulation of mud or slush in the wheel covers, undercarriage bays, including the leading edge, and underside of the wing and tail plane.
- Check that all openings are free of ice, snow and mud. These include the crankcase breather, pitot tube, static ports, fuel vents, heater intake, carburetor intake, control cable openings, turbo compressor inlets and wheel wells.
- Remove all ice, snow and frost from the airplane lift and control surfaces (even a little frost can significantly affect the lift). Never assume a little ice or frost is acceptable, and do not rely on snow blowing off during takeoff. Loose snow can be swept off with a broom, while ice can be removed using de-icing fluids or placing the airplane in a heated hanger. If the latter method is employed, make sure the water does not run into the control surface hinges or crevices, with the result that it freezes when the airplane is taken outside. Alternatively, go later in the day when the sun has melted any snow or ice.
- Brakes can freeze after landing and parking the airplane. Prior to starting the engine, check whether the brakes are frozen or not, by moving the airplane using a tow bar.
- Be aware that when an airplane is pulled from a warm hanger into falling snow at near freezing temperatures, snow will melt on contact with the warm airplane. Subsequently, an ice jacket can form to which snow adheres tenaciously and will not blow off during takeoff.
- Make sure any de-ice and anti-ice equipment is functioning properly (e.g. pitot heat, carburetor heat, alternate air and boots).

Engine Starting

Cold engine starts require familiarity with the manufacturer's cold starting procedure and the associated hazards (e.g. engine damage, over priming, freezing spark plug electrodes and engine fires). Preheat might well be needed, with extra caution being taken because of the fire risk, depending on the method used. The easiest way is to leave the airplane in a heated hanger overnight.

After engine start, carburetor heat may be needed during taxiing to enrich the mixture and help vaporize the fuel. Allow adequate time for the engine to warm up before performing the run-up checks. On the other hand, avoid prolonged idling, as spark plugs may become fouled due to insufficient heat generation.

Taxiing

There are a number of hazards associated with taxiing an airplane depending on the conditions.

> ▸ Braking effectiveness is poor on snow and ice.
> ▸ Snow or slush melting on warm brakes as you taxi, can freeze when you stop (e.g. for the run up) resulting in locked brakes.
> ▸ Snow, slush or water thrown up on the wheels, wheel wells or control surfaces, can freeze, resulting in locked controls or landing gear.
> ▸ The airplane can slide if the run-up is not carried out on a dry location.
> ▸ Conditions can be very slippery (you can slide sideways in cross wind conditions).
> ▸ Snow may obscure obstacles, ruts and puddles.
> ▸ Steering control may be less effective on snow and ice.
> ▸ Windscreens can readily mist over due to moisture from damp clothing and footwear. A handy cloth can be useful until the de-mister functions adequately.

Awareness of these hazards, as well as proficiency at handling such conditions, is required. Taxi slowly and avoid areas of water and slush.

Takeoff

Pitot heat should be turned on well before takeoff and adequate time allowed during taxiing for the gyro instruments to spin up from their cold condition. It may be appropriate to inspect the runway prior to takeoff, so that any potential hazards are identified.

Note that snow, slush, mud or wet grass will lengthen the takeoff roll or, prevent takeoff altogether. Allow generous safety margins. Also ensure the condition of the full length of the runway is known, as the slowing effect due to snow, slush and water, is most serious towards the end of the takeoff roll.

If the airplane has been parked outside overnight, ensure the heater and de-mist vents have been purged of moist air prior to takeoff. There have been a number of instances of the windscreen completely misting over on takeoff on a VFR day.

En-Route

While winter weather can result in beautiful clear days, the weather, much of the time, can be overcast and changeable. This, coupled with the cold temperatures and snow covered terrain, results in a number of additional hazards.

> - The likelihood of carburetor icing and airframe icing.
> - A possibility of rapidly deteriorating visibility caused by snow or rain. Visibility can also deteriorate slowly. There is a danger that the pilot continues with the flight in the hope that the visibility will improve.
> - Terrain that looks very different when covered with snow, resulting in familiar landmarks being masked.
> - Differing weather from forecast (e.g. wind strength/direction, snow showers).
> - Whiteout that occurs when snow covered featureless terrain blends into an overcast sky. (The phenomenon can occur in clear weather, thus a pilot should be prepared.)
> - A lack of weather reporting stations in remote locations.
> - An altimeter reading several hundred feet high because of very cold temperatures.

Extra vigilance and awareness of the possibility of these hazards is therefore necessary, together with a readiness to take avoidance action as required. Proficiency at instrument flying is recommended when flying cross-country in less than ideal conditions.

Descent and Landing

During descent, sufficient power should be maintained to keep the engine warm. This can result in the descent profile being extended.

Landing surfaces in cold climates can be very treacherous and conditions can change rapidly. Therefore, up-to-date information should always be used (if available). Hazards include:

> - Ice fog.
> - Blowing snow.
> - Difficulty in depth perception on a dull day, caused by the lack of shadows.
> - An inability to properly see the runway markings and lights.
> - Snow-drifts on the runway.
> - The presence of snow banks on the edge of runways.
> - Poor braking effectiveness and directional control.
> - Possibile frozen brakes resulting in directional control difficulties on landing.
> - Equipment on the runway that may not be readily visible.
> - A runway where the full length may not have been cleared of snow.
> - Runways covered in snow, ice, slush or water, which will increase the landing distance required (as breaking effectiveness is reduced), and make control difficult.

Winter flying can be most enjoyable, but there are many hazards. Preparation, awareness, proficiency and careful decision making can minimize the risks.

Avoiding and Dealing with Icing

7

Icing can take two forms - *Induction System Icing*, which affects engine performance, and *Airframe Icing*, which affect the aerodynamics and instruments of the airplane. Both can have serious consequences, and therefore knowledge about avoidance, and how to deal with icing, should it occur, is necessary for safe flying.

Induction System Icing

Induction system icing is a well-known phenomenon, but continues to be a probable cause of a number of accidents or incidents involving engine power loss. It involves any ice formation in the induction system of airplanes equipped with reciprocating engines. It forms subtly and insidiously.

Types of Induction System Icing

There are basically two types of induction system icing, namely carburetor icing and impact icing.

Carburetor Icing (or Refrigeration Icing)

Carburetor icing is caused by the venturi refrigeration effect in the vicinity of the butterfly or throttle plate (throttle ice) or, when water, held in suspension, precipitates and freezes in the induction piping (fuel ice). Throttle and fuel icing occurs in conventional float type carburetors, and to a lesser extent in pressure jet type carburetors. It is not an issue with fuel-injected engines.

Impact Icing (or Intake Icing)

Impact icing is formed by the striking of visible moisture-laden air (rain, snow, sleet and cloud) at temperatures below freezing, on elements of the induction system, which are at temperatures of zero degrees or lower. It can affect the air intake, air filter, carburetor heat selector valve and alternate air selector valve. Impact icing occurs in conventional carburetor systems, as well as fuel-injected systems.

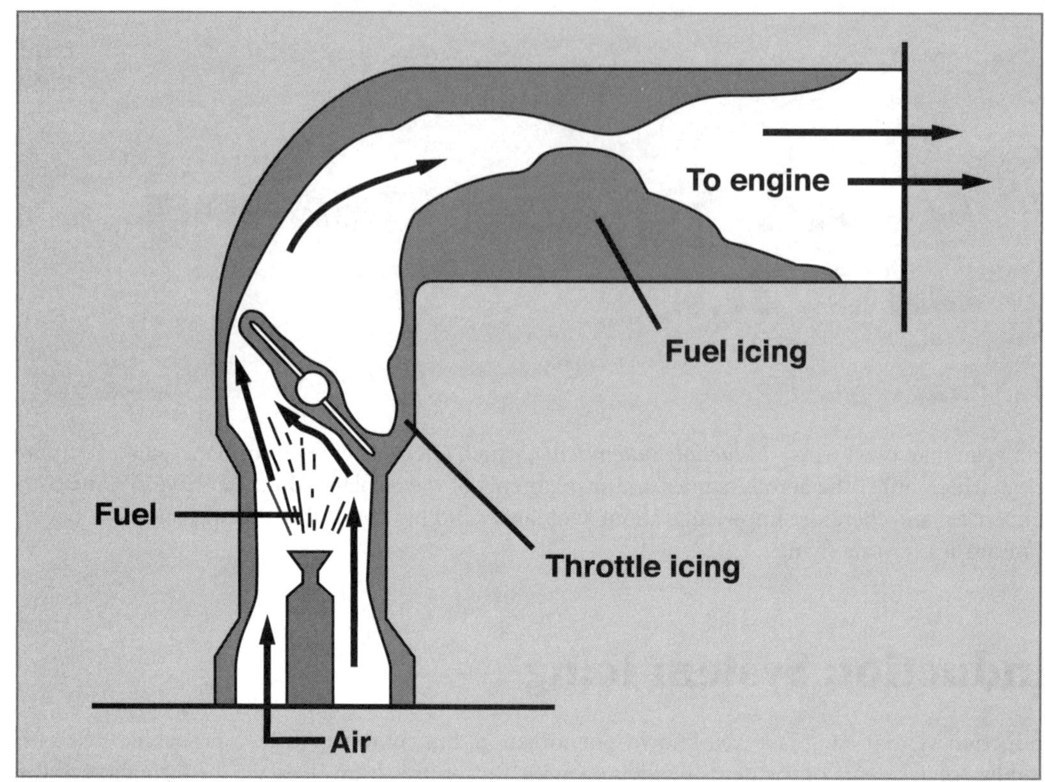

Both effects can cause serious loss of fuel/air mixture flow to the engine. Therefore, positive action (using carburetor heat or alternate air) must be taken if there is a high probability of icing, or at the first warning sign of icing.

Likelihood of Carburetor Icing

Some reciprocating engines are more prone to carburetor icing than others, depending on the level of radiant heat and the type of carburetor. Carburetor icing is a function of dew point and the ambient temperature. Probability charts are widely published.

Carburetor icing can form any time there is moisture in the air (i.e. humidity in clear air, or visible moisture such as rain, cloud, drizzle or sleet) at temperatures between -10°C and 38°C, and can occur at any power setting. Below -10°C entrained moisture forms ice crystals, and therefore passes harmlessly through the induction system. Carburetor icing is most likely to occur:

- ▸ At low power settings when the partially closed butterfly plate creates a greater venturi cooling effect (e.g. during descent and taxiing).
- ▸ In the range -5°C to 20°C when the relative humidity is above 50%.

Other warning signs include:

> ‣ Poor low level and surface visibility (early morning and late evening).
> ‣ Wet ground and minimal wind.
> ‣ Flying just below the cloud base (high humidity).
> ‣ In clear air, just after fog or cloud has dispersed.

The possibility of carburetor icing increases as the temperature decreases and as the relative humidity increases. It is also more probable at dusk, when the air cools and becomes more humid, as well as at higher altitudes as the air gets cooler and humidity increases. However, carburetor icing is not restricted to any one season and it should not be assumed, that because it is a "nice day", you will not get carburetor icing.

The probability chart below is easily used with knowledge of the temperature and dew point, which can be obtained from weather reports.

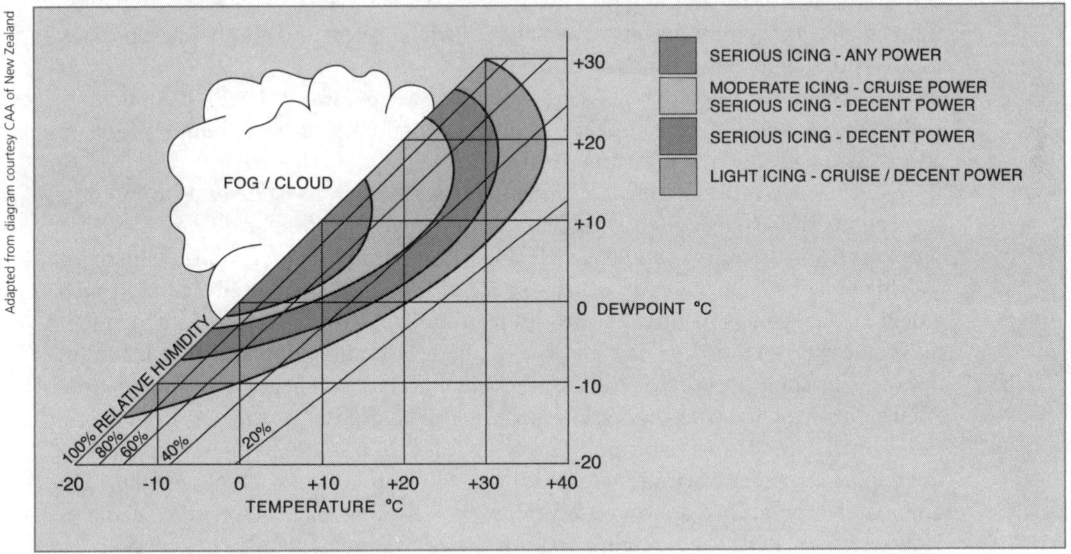

Warning Signs of Carburetor Icing

Pilots need to be vigilant for the warning signs of carburetor icing as indicated by:

> ‣ Loss of power, indicated by a slow decrease in engine rpm (fixed pitch propeller) or a slow decrease in manifold pressure (variable pitch propeller). Loss of power may also be detected by a decrease in airspeed or altitude (e.g. noticing need for nose up trim).
> ‣ Unexplained engine roughness.
> ‣ Slowly decreasing exhaust gas temperature (EGT) (with constant power setting).
> ‣ Possible back firing.

Preventing Carburetor Icing

It is better to prevent ice build-up, than to attempt to melt it. Preventative measures include:

- During preflight planning, obtain the atmospheric temperature and dew point to ascertain the likelihood of carburetor icing.
- Know the operation and function of the airplane carburetor heat or alternate air systems.
- Check the carburetor heat function and effectiveness during the engine run up, noting the "heat on" power drop. Keep heat on for 30 seconds if no guidance is provided by the airframe manufacturer. If ice is present, the application of carburetor heat results in a drop in rpm or manifold pressure, followed by a gradual increase. When the heat is then turned off, the rpm/manifold pressure will read higher than the starting point.
- If there is a long taxi, or a delay when high-probability icing conditions exist, apply carburetor heat at the holding point to remove any ice that may have accumulated.
- Remain alert for symptoms of ice formation during takeoff and climb out, especially in high probability icing conditions.
- If the airplane is fitted with a carburetor air temperature (CAT) gauge, apply sufficient carburetor heat during cruise to maintain a safe temperature in accordance with the manufacturer's instructions.
- If no CAT gauge is fitted, the recommended procedure, in most cases, is to monitor the engine rpm (fixed pitch propeller) or the manifold pressure (variable pitch propeller), and as soon as a decrease is noted, apply full carburetor heat. The onset of icing must be detected early, otherwise ice formation may have advanced to such a degree that there is insufficient heat to melt the ice (carburetor heat is dependent on continued operation of the engine). If there is a high probability of icing, an option is to apply carburetor heat at frequent intervals to check for ice formation. Alternatively, if icing is severe, carburetor heat can be used as an anti-ice mechanism, rather than a de-ice mechanism, by applying full carburetor heat continuously with the engine leaned and with power less than 75%. Full heat at high power and high ambient temperature can cause cylinder over heating and detonation damage. However, the POH and engine manufacturer's manual should be checked to determine the correct procedure for the airplane type. Remember that the use of carburetor heat enriches the mixture unless it is adjusted to compensate.
- Prior to closed throttle operation (e.g. descents), or operation below 2,000 rpm, apply carburetor heat before closing the throttle, and leave it on while operating at reduced power. It is preferable to have some power on during long descents to maintain a warm engine. Alternatively, warm the engine every 500 feet during a prolonged descent.

Remedial Action if Carburetor Icing is Suspected

If carburetor icing is suspected, the immediate action is to eliminate the ice which has already formed downstream of the carburetor heat inlet. The POH procedure for carburetor icing should be followed, with the application of full carburetor heat. Ensure that heat is left on for a sufficient length of time to melt any ice (this can take up to several minutes). The engine may run rough, and even momentarily cut out, as a piece of ice is melted and water, or partially melted ice, is ingested through the system. You may need to lean the mixture to compensate for the less dense air and restricted airflow. Leaning the mixture has the added benefit of obtaining higher temperatures from the engine. Shutting off the carburetor heat too early will result in more ice build-up and the possibility of complete engine failure.

Likelihood of Impact Icing

Impact ice is formed between -15°C and 0°C when visible moisture is present (rain, snow, sleet, drizzle, cloud). Build-up is most rapid at -5 °C.

Remedial Action if Impact Icing is Suspected

The warning signs for impact icing are the same as for carburetor icing, except in some fuel-injected engines, where the alternate air source operates automatically. In this case, the only indication may be airframe icing.

Where alternate air is not automatic, the remedial action is the same as for carburetor icing. By applying carburetor heat or alternate air, the blockage is by-passed and the effect on engine performance will be immediate (provided this is done early, as there is the potential for the selector valve to freeze if action is taken too late). However, the air must remain on until the airplane is clear of impact icing conditions and the ice has melted.

Airframe Icing

Airplane airframe icing is a known danger to all IFR pilots. It occurs when visible moisture (e.g. cloud, rain and drizzle) comes into contact with parts of the airframe that are below the freezing temperature. Even small amounts of ice can seriously affect lift and drag. Without proper ice protection equipment, you are not permitted to fly into known icing conditions. It is therefore important to adopt avoidance strategies, as well as know what action to take should icing be inadvertently encountered. Remember that icing is difficult to forecast and can be localized. Even if you have de-icing equipment, there are limitations, and it is prudent to only stay in icing conditions for the minimum amount of time. The most effective strategy, therefore, is avoidance.

Likelihood of Airframe Icing

The likelihood of airframe icing depends on the air temperature, the amount of water present and the size of the water droplets. If you operate an airplane in cloud between 0°C and -20°C, you can expect to accumulate some form of ice.

Main Types of Airframe Icing

There are three main types of airframe icing – clear ice, rime ice and mixed ice.

Rime Ice

Rime ice is commonly formed in stratiform cloud at low temperatures (-10°C to -20°C) when small droplets freeze immediately on contacting the airplane surface. It is a porous, white, granular brittle (frost like) coating and adheres to the leading edge of wings or struts as well as the windshield. It has little tendency to spread and can easily be removed by airplane de-icing systems.

Clear Ice

Clear ice is normally formed in cumuloform cloud at temperatures just below freezing (0°C to -10°C) when large droplets or freezing rain spread over a surface. It is hard, glassy, transparent or translucent and difficult to see. It can change the shape of the airfoil and is difficult to remove. It is the most dangerous form of icing.   Freezing rain is the most severe form of clear icing. It occurs in conditions where "warm" moist air is moving into a cold region and overruns the colder layer beneath. Clear ice can also form when a "cold soaked" wing (with integral fuel tanks) enters rain.

Mixed Ice

Mixed ice is a mixture of rime and clear ice, and forms at temperatures between -10°C and -15°C.

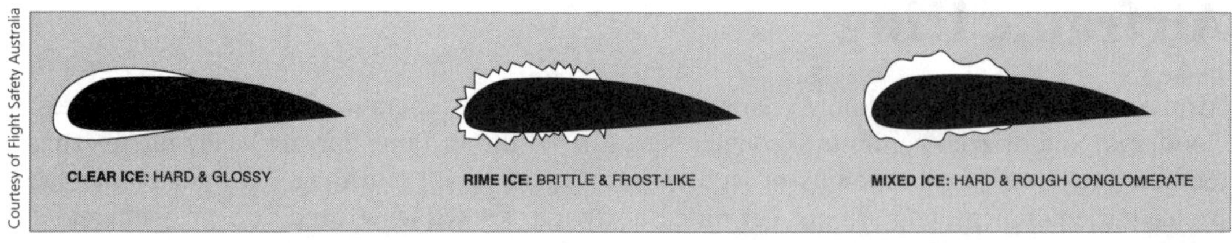

Effects of Airframe Icing

The most hazardous ramification of airframe icing is the effect on the aerodynamic performance of the airplane, through changes in the shape of the airfoil. This has the effect of:

> - Increasing the stall speed.
> - Reducing the stalling angle of attack.
> - Increasing the amount of drag.
> - Decreasing the maximum lift.
> - Adversely affecting the airplane handling qualities.

Note that ice buildup is not just restricted to the airfoils. It also accumulates on all exposed frontal areas, including the windshield, antennas, vents and intakes. Additional hazards therefore include:

> - A reduction in engine power due to the possible formation of impact icing.
> - Accumulation of ice on the control surfaces, resulting in degraded or complete loss of movement.
> - Additional airplane weight due to accumulation of ice on exposed airframe surfaces.
> - Blockage of the air filter.
> - Deterioration in the effectiveness of the trim(s).
> - Reduced propeller efficiency.
> - Asymmetric shuddering, due to propeller ice shedding.
> - Possible damage to wing flaps during extension/retraction.
> - Possible freezing of the landing gear mechanism.
> - Obstruction of vision.
> - Damaged or degraded communications and navigation equipment.
> - Possible pitot tube and static air blockage.
> - Possible instrument malfunction (e.g. airspeed indicator and altimeter).
> - Blockage of the fuel vent that can result in fuel starvation.
> - Possible uncontrolled roll phenomenon (roll upset).
> - Possible tailplane stall.
> - Reduced engine-cooling effectiveness.

Preflight Planning Actions to Avoid Airframe Icing

Preparation is the key strategy for avoiding icing and managing any inadvertent icing encounter. It should include an assessment of your knowledge, experience and proficiency with respect to flying in conditions conducive to airframe icing.

- Obtain a comprehensive weather forecast and briefing. Check:
    - For any forecast icing on takeoff, en-route and landing.
    - The extent of frontal activity, convective activity and rising terrain when below freezing temperatures are forecast aloft (this will aid in ascertaining the likelihood of localized icing).
    - The cloud tops and thickness.
    - The locations of any areas of warm air.
    - The likelihood of freezing rain.
- Avoid the "freezing zone" (0°C to -20°C) when there is visible moisture or the likelihood of freezing rain.
- Pick a route where the minimum en-route altitude is below the forecast freezing level or you are able to fly in clear conditions.
- Avoid flying just above the tops of cloud as they can rise rapidly and water concentration is greatest near the top of cloud.
- Have an escape plan should icing be encountered en-route.
- If you have anti-icing equipment, make sure it is functioning properly before takeoff, and that you plan for minimum exposure during climb and descent.
- Provide extra fuel margin.
- Review the POH to refresh your memory on the airplane icing limitation and procedures.

In-flight Precautions

If flying in conditions where airframe icing is possible, then the following actions should be taken:

- Request weather updates and pilot reports.
- Consider climbing through possible icing conditions at a higher airspeed than normal.
- Switch on the pitot heat and any other electrical anti-icing equipment (e.g. propeller heat) well before entering icing conditions to give the equipment time to warm up.
- Switch on the windshield defroster.
- Monitor the outside air temperature gauge.
- Regularly check for ice accumulation, particularly on small protrusions where it generally forms first. Also monitor the airspeed as a possible indication of ice accumulation.
- If the airplane has de-icing boots, then they can be cycled from time to time to prevent the valves in the pneumatic system from sticking. Check the POH for the procedure relative to the airplane being flown.

Actions if Icing Conditions are Inadvertently Encountered

Always first look for ways to avoid icing, such as diverting to remain clear of cloud or flying on top. If this is not possible, then the most important thing is to detect the formation of ice at the earliest possible stage. Note that if you can see icing on the wing, you will already have accumulated some ice, with the consequent increase in airplane weight. If icing is encountered the following actions are generally appropriate, but the POH should be checked to determine the procedures and limitations specific to your airplane.

> ▸ At the first sign of icing, act promptly.
> ▸ Switch on the pitot heat, carburetor heat and any other anti-icing equipment. (They should already be on any time you are flying in cloud or precipitation.)
> ▸ Activate any de-icing equipment in accordance with the relevant manufacturer's instructions.
> ▸ Take immediate action to implement your escape plan (do not allow the ice to accumulate). Either:
>> ▸ Descend to a lower (safe) altitude that is warmer and/or free of cloud.
>> ▸ Climb to cruise on top of cloud, provided you are certain it is not too thick.
>> ▸ Divert to an area that is warmer and/or free of cloud.
>> ▸ Go back.
> ▸ Turn off the autopilot, as this can mask the aerodynamic effect of ice.
> ▸ Do not extend the flaps.
> ▸ Avoid abrupt maneuvers.
> ▸ Reduce the angle of attack by increasing air speed.
> ▸ Select alternate engine intake air (if not automatic).
> ▸ Advise ATC, and seek help if necessary. (If the situation is serious, do not hesitate to declare an emergency, divert and land as soon as possible.)
> ▸ If you have to land an iced-up airplane, fly with extra airspeed (but not too much) with zero flap. (While higher airspeed reduces the angle of attack of the wing and therefore reduces the likelihood of a stall, it increases the angle of attack of the tailplane (as does the use of flaps) thus increasing the risk of a tailplane stall.)

Airframe icing effects can be unpredictable, and depend on a wide range of variables. Its formation can create a situation from which the pilot may have difficulty recovering and, in some cases, may not recover at all. The best strategy is avoidance.

Takeoff and Landing Risk Minimization

8

Takeoff and landing are the most *critical stages of flight* – low speed and low to the ground.

Courtesy of Australian Transport Safety Bureau

"The airplane was significantly heavier than the pilot had calculated and the strip length available was insufficient for the intended operation, were two factors that may have contributed to this accident"

Statistics worldwide show that a large proportion of all general aviation accidents occur during the takeoff, approach and landing phases of flight. Most are due to lack of proficiency and poor judgment, with macho attitudes, such as ego and loss of face, also playing a part.

While many of these accidents are weather related, others have been due to:

> ▸ Inadequate takeoff and landing performance assessment.
> ▸ Lack of understanding of all the risk factors that are applicable to takeoff and landing.
> ▸ Flying an overloaded or out of balance airplane.
> ▸ Forgetting to lower the landing gear.
> ▸ Inadequate assessment of the risks associated with night takeoff and landing.
> ▸ Inadequate assessment of bush airstrips.

This section discusses these issues and the associated avoidance strategies and tactics.

Takeoff and Landing Performance Assessment

When taking off or landing, the performance of the airplane must be known to ensure that the available runway distance is adequate and that the required climb gradients are achievable. Under normal circumstances the runway distances and obstacles are such that there is ample margin. However, accident statistics show, that when operating from airfields that have a short runway, a surface other than hard paving, are at high density altitudes or have significant obstructions, there have been numerous cases of airplanes:

> ▸ Failing to get airborne in the distance available.
> ▸ Overrunning the end of the runway on landing.
> ▸ Colliding with obstacles, after failing to achieve the required climb gradient.

These accidents are generally the result of inadequate takeoff and landing performance assessment which requires, in the first instance, an understanding of the variables that affect takeoff and landing performance, and secondly, an ability to determine the airplane's performance on the basis of the prevailing conditions and pilot proficiency.

Variables Affecting Takeoff and Landing Performance

Numerous variables affect the takeoff and landing performance of airplanes:

Atmospheric Conditions

> ▸ Airfield altitude, air pressure and temperature.
> ▸ Wind direction and speed (headwind, tailwind, crosswind and gusting wind).
> ▸ Gradient wind.
> ▸ Humidity.
> ▸ Turbulence and wind shear.

Runway Characteristics

- Type of runway surface (bitumen, concrete, long grass, short grass, green grass, dry grass, gravel, soil, hard packed sand, clay).
- Runway condition (extent of free water, moisture content, degree of compaction, extent of irregularities, sub surface softness).
- Slope of the runway (longitudinal and transverse).

Airplane Operating Characteristics

- A pilot's takeoff and landing technique and skill.
- Takeoff or landing gross weight of the airplane.
- Takeoff or landing speeds.
- Flap setting.
- Engine efficiency (age, condition).
- Propeller efficiency.
- Inadvertent braking during takeoff.
- Tire pressures.
- The rate of acceleration.
- Clogged wheel fairing(s).

These variables affect the airplane aerodynamic performance, engine performance, rolling resistance or braking effectiveness.

Aerodynamic and engine performance decrease with an increase in pressure altitude and temperature, while humidity decreases engine output. The result is, an increase in takeoff and landing distances and lower climb gradients for a given airplane weight. Any increase in weight reduces the aerodynamic performance.

The rolling resistance is a function of tire pressure, airplane weight and the surface characteristics of the runway. Generally, the biggest influence is the runway surface and its condition. While the rolling resistance on hard pavement is minimal and predictable, the resistance of other surfaces varies widely, and is far more difficult to assess.

Landing on grass can result in an increased ground roll, despite increased rolling resistance. This is because braking effectiveness is reduced due to lack of tire friction. Short dry grass can be very slippery.

Takeoff and landing distances can also vary by more than 20% due to differences in pilot technique and skills.

Takeoff and Landing Performance Charts and Tables

Regulations require pilots to assess the airplane performance under the prevailing conditions in order to assure the safety of the airplane during takeoff or landing. Takeoff and landing charts and tables can be found in the airplane POH, and permit a pilot to:

> ▸ Determine the required takeoff or landing distance for the main variables affecting takeoff and landing (i.e. weight, temperature and pressure altitude).
> ▸ Determine climb gradients to check that all obstacles can be cleared.
> ▸ Determine any weight restrictions that may have to be imposed because of a runway length limitation and/or the climb gradient requirements.

This data has, however, a number of limitations:

> ▸ It is based on new airplanes flown by highly experienced pilots under ideal and exact conditions. The actual takeoff and landing performance by the normal pilot will invariably be different (e.g. longer roll, different speeds, higher round out).
> ▸ It generally does not account for some less common variables affecting takeoff and landing performance. In particular, runway characteristics other than flat hard paving.

The distances determined using the POH data may therefore have to be adjusted to account for those conditions not covered in the manual and a factor of safety applied to address any uncertainties.

Adjusting Takeoff and Landing Distances Factors

The following table gives some generally accepted factors that can be applied to takeoff and landing distances determined using the POH charts and tables. Their application should be treated with caution and judgment, as they are approximations only, covering a range of light airplanes.

Condition	**Takeoff** - *distance increase to height of 50 feet*	**Landing** - *distance increase from height of 50 feet*	**Factor**
A 2% slope *(See note 1.)*	Uphill 10%	Downhill 10%	1.1
Dry grass *(up to 20 cm on firm soil - See note 2.)*	20%+	20%+	1.2
Wet grass *(up to 20 cm on firm soil - See note 3.)*	30%	30%+	1.3
Soft ground or snow. *(See note 4.)*	25%+	25%+ Light snow or surface muddiness	1.25+

All the factors are cumulative. That is, when several factors are relevant, they must be multiplied.

1. Source: CAA UK AIC 12/1996. 2. Source: CAA UK AIC 12/1996. If the grass is very short it can be extremely slippery and the landing distance can increase by up to 60%. Long wet grass can increase the distance by up to 40%. 3. Source: CAA UK AIC 12/1996. 4. Source: CAA UK AIC 12/1996/CASA AC 91 - 225. The effect of soft ground and snow is variable and possibly unpredictable. Extreme caution should therefore be exercised.

Providing a Factor of Safety

No adjustment factors can reasonably cater for all variables affecting takeoff and landing. Nor is it possible to assess with accuracy the exact nature of some of the variables that may be applicable. A factor of safety should therefore be applied to address:

> ▸ Performance variables not covered by any adjustment factors (e.g. pilot technique, flight parameters, engine condition).
> ▸ The approximate nature of the adjustment factors.
> ▸ Any uncertainties with respect to runway characteristics, atmospheric conditions, airplane operating characteristics or pilot skill.

The factor of safety selected will depend on the circumstances and any regulatory requirements. "*The Safe Pilot's 12 Golden Rules*", published some years ago by the National Aviation Underwriters, suggested a factor of 1.8 if the runway is a hard surface, a factor of 2.0 if sod, and 3.0 if wet grass. Other publications suggest different factors and therefore every situation needs to be carefully assessed, and an appropriate safety factor applied. As a means of checking your margin, you should always nominate a decision point where you can safely abandon the takeoff or discontinue the approach if things are not going as expected. A well-known rule of thumb for takeoff is to note the halfway point of the runway, and if you have not achieved lift off speed by that point, abort the takeoff. For landing, the decision point is the point at which there is sufficient room to effect a safe go-around. Always take into account the go-around climb capability of the airplane and the nature of obstacles at the end of the runway, when determining the landing distance required.

Minimizing Takeoff and Landing Risk Factors

While the determination of the takeoff and landing performance of an airplane, limits the risks of incurring an accident, it is wise to not solely rely on the factor of safety to cover all potential risks. The best strategy is to consider all the possible risk factors, relevant to the particular takeoff or landing, and apply tactics wherever possible, that reduce these risks to an absolute minimum. Consider the following:

Airstrip Information

> ▸ Always obtain the current information about any airfields you plan to use (or may use). Pay particular attention to bush airstrips.
> ▸ If you are at an airfield and don't know the length of the strip, step out the runway using 2.5 feet (approximately 0.75 meters) per step.
> ▸ If an airstrip has an unknown slope, taxi the airplane to both ends and note the altimeter difference.
> ▸ Walk the strip if the runway conditions are not obvious.
> ▸ The wind speed can be estimated from the windsock angle, or other indications if there is no windsock.

Safe Practices

- Never fly an overloaded or out of balance airplane.
- Fly early in the day to take advantage of the cooler temperature.
- At altitudes above 5,000 feet, a normally aspirated engine may need to be leaned for takeoff in accordance with the POH.
- Prior to landing at an unlicensed bush airstrip, inspect the field from the air and check out the approach. The precautionary landing procedure outlined in Section 14 can be used if appropriate.
- Ensure that the specified POH takeoff and landing technique, and airplane configuration, is used when operating at airfields where distance and obstacle clearance are factors.
- Always commence a takeoff from as close as possible to the approach threshold.
- If your proficiency at a crosswind takeoff or landing is deficient, consider your options carefully if such a maneuver is required.
- If you are not fully proficient at a short field takeoff or landing, provide extra margin.
- The decision to takeoff uphill and into wind, or downhill and downwind, is a trade off between wind and slope. Some points to note under these circumstances include:
    - If you takeoff downhill and downwind, your climb gradient will reduce as you gain altitude because of the normal wind gradient.
    - Ensure that the climb gradient is sufficient to out climb any rising terrain.
    - If taking off towards the lee side of hills, down drafts can result in a decreased climb gradient.
- The decision to land uphill and downwind, or downhill into wind, is again, a trade off between wind and slope. Consideration should also be given to any airplane limitations noted in the POH, as well as the inherent risks of a downwind landing. In making the assessment, the following should be noted:
    - When landing downhill an extended flare usually results. This will increase the landing distance required.
    - A go-around is always a possibility, and care should be taken to ensure that the airplane is able to out climb any obstacles or rising terrain.
    - Be careful of visual illusions associated with landing.
    - When landing downhill, braking effectiveness is reduced.
- Ensure that any obstacles can be cleared by at least 50 feet.
- If the wind at ground level is negligible, the direction of landing or takeoff should be into the gradient wind. Failure to do this can impact on threshold landing speed or takeoff climb gradient. Gradient wind on landing can be determined by cross wind drift, and on takeoff, by the area forecast.
- Note that gusty conditions require an increase in speed (by one half the Gust Factor, up to 10 knots), which in turn requires a longer takeoff or landing roll.
- Avoid "runway fixation" or "threshold fixation".
- Always be ready for turbulence and its effect on the approach.
- If there is any doubt about making a safe touch down, take the airplane around – it is one of the best accident avoidance procedures available. Don't try to make a bad landing good, and avoid "pilot indecision".

Conditions and Situations to Watch

- Be aware of the risks associated with night takeoff and landing (e.g. somatogravic illusion, black hole effect, ground light confusion and other illusions).
- Avoid very long grass, particularly with low wing airplanes, as the pitot tube and air filter can potentially get clogged. In addition, it is difficult to detect runway obstructions.
- A soft field takeoff, in combination with a short field, is a very high-risk proposition.
- Be careful with short strips and obstructions when landing. If there is a need to go around, the obstacle may not be cleared if the decision is left too late.
- Low level wind shear.
- Gusting conditions can cause any crosswind to potentially exceed the POH limit as well as cause possible over reaction by the pilot, depending on the level of proficiency.
- Be aware of the hazardous weather warning signs that can affect takeoff and landing
- Frost can turn a grass strip into a skating rink.
- When taking off on gravel, the power is normally applied slowly to avoid stone chip damage to the propeller. A longer takeoff run will therefore be required.
- Asymmetric braking, rudder deflection, or nose wheel steering may be required to handle transverse slopes, and consequently a longer runway distance will be required.
- Grass resistance varies with the length, freshness, moisture content and density.
- Free water can, at low speeds, increase resistance because of a build-up of water in front of the tires. However, at high speeds this can result in aquaplaning.
- Unpacked snow or slush can double the takeoff distance.
- Grassed runways that look dry can be very soft underneath due to recent rain.
- Note that the length of the runway required for landing is generally less than that required for takeoff. Hence, while you may be able to land, you may not be able to takeoff.
- Some airstrips have a camber and, in a crosswind, the combination could result in the airplane veering off the runway. The remedy is to land on the upwind side of the runway.

Safety conscious pilots know, and are able to recognize, the factors that affect takeoff and landing performance. They also know, that while takeoff and landing performance can be calculated for most airfields, there may be instances where the conditions are uncertain. Under these circumstances, cautious assessment and prudent judgment will result in the risks being minimized. Failure to appreciate the airfield conditions and the performance limitations of the airplane and pilot, is a high-risk proposition.

Wake Turbulence

Wake turbulence is a well-known phenomenon. It occurs as the result of wake vortices with the strongest vortices produced by heavy clean airplanes flying slowly.

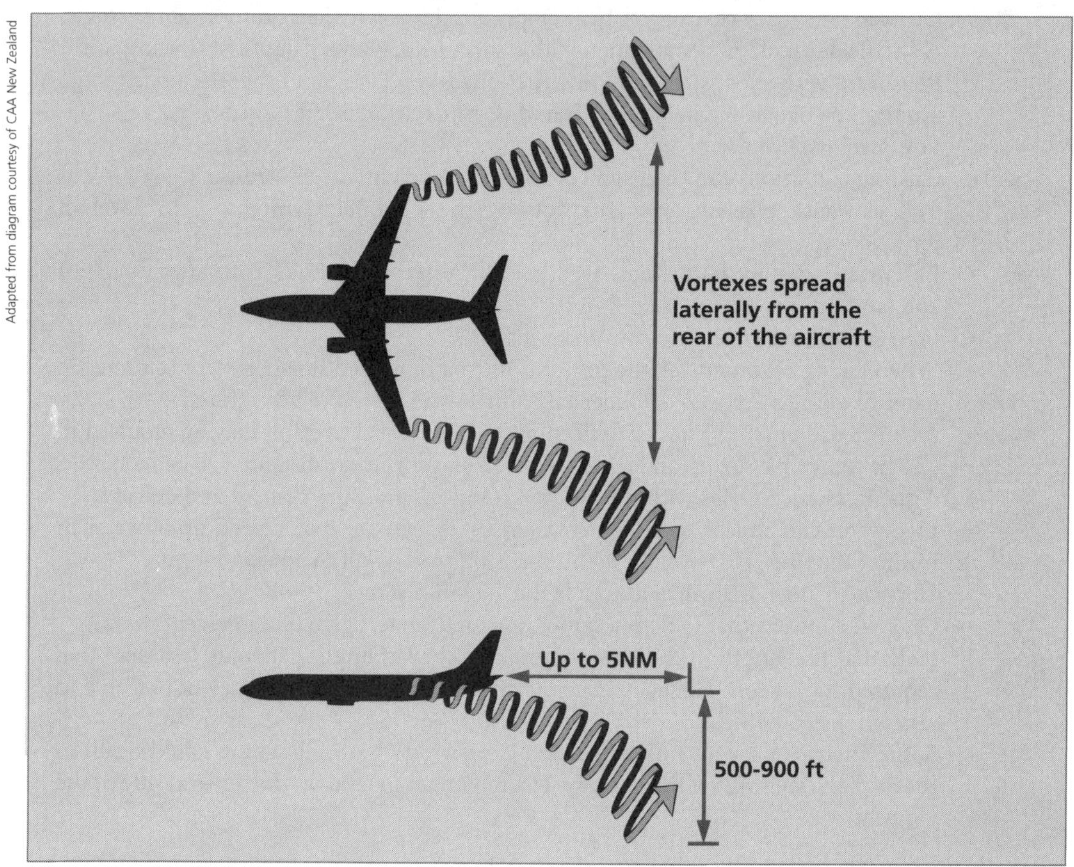

While many private pilots do not fly in airspace containing nearby heavy airplanes, it is wise to be aware of the general guidelines for avoiding wake turbulence in case the situation is ever encountered.

> ▸ Wait a few minutes before requesting a clearance to cross a live runway when a large airplane has just taken off or landed. Also, when holding near a live runway, expect wake turbulence.
>
> ▸ During takeoff, if you consider wake turbulence from a preceding airplane may be a problem, wait 2 to 3 minutes before taking off. When departing behind a larger airplane, plan to lift off before the takeoff rotation point of the preceding airplane and stay above its flight path. Try to turn off the departure path as soon as possible in case you cannot out climb the larger airplane. If taking off following the landing of a large airplane, plan to become airborne after the point of touch down of the landing airplane.

> ▸ When flying behind a large airplane landing on the same runway, stay at or above the preceding airplane's final approach flight path, note the touch down point and plan to land beyond that point. When landing after a large airplane on parallel runways, beware of possible drifting of vortices onto your runway. When landing behind a departing airplane, plan to touch down well prior to reaching the rotation point of the departing airplane. When landing behind another airplane on a crossing approach, cross above the airplane's flight path or abort the landing.
>
> ▸ Note that crosswinds affect the position of vortices. Adjust the takeoff and landing paths accordingly.

In short – takeoff before, land after and stay above a preceding larger airplane.

Weight and Balance

Overloading and/or exceeding the center of gravity limits have been contributing causes of takeoff accidents, with the risk being aggravated when airplane performance reducing factors are present (e.g. high density altitude). It is therefore not always possible to fill every passenger seat, load luggage to capacity, or carry full fuel. In addition, care must be taken to ensure that the load distribution is such that the center of gravity limits are not exceeded.

Effects of the Weight and Balance Envelope Being Exceeded

The consequences of exceeding the weight and balance envelope of an airplane are considerable.

Effects of Overloading an Airplane

- ▸ Slower acceleration.
- ▸ Higher takeoff speed, leading to an increased takeoff run.
- ▸ A decreased angle of climb, resulting in reduced ability to clear obstacles.
- ▸ Impaired maneuverability.
- ▸ Impaired controllability.
- ▸ Increased stall speed.
- ▸ Reduced structural strength margins.
- ▸ Reduced maximum altitude capability.
- ▸ Reduced rate of climb.
- ▸ Reduced single engine climb performance on twin-engine airplanes.
- ▸ Higher landing speed, resulting in a longer landing distance.
- ▸ Reduced braking effectiveness.
- ▸ Higher fuel consumption.
- ▸ Reduced range.

Effects of Exceeding the Forward Center of Gravity Limit

▶ Difficulty in rotating to takeoff attitude.
▶ Increased stall speed.
▶ Higher fuel consumption.
▶ Difficulty in flying a stable approach (inadequate nose up trim).
▶ Difficulty in flaring and holding the nose-wheel off.
▶ Increased load on the nose-wheel landing gear.

Effects of Exceeding the Rear Center of Gravity Limit

▶ Premature rotation, or inadvertent stall, in the climb or go-around.
▶ Difficulty in trimming, especially at high power.
▶ Decreased longitudinal stability, particularly in turbulence.
▶ Degraded stall qualities to an unknown degree.
▶ Difficult or impossible spin recovery.

The more aft the center of gravity, the more unstable the airplane (with particularly violent stall characteristics). Note that if the all up weight limit is exceeded, the center of gravity limits are generally invalid.

Eliminating Weight and Balance Risk

It is not hard to remain within the weight and balance envelope, provided some precautions are taken.

▶ Always carry out a weight and balance check when you are operating near full gross weight, or are carrying passengers or baggage that might involve abnormal loading.
▶ It may not always be possible to take the maximum number of passengers, luggage or fuel. There are many four-place, and six-place, airplanes where the fuel tanks cannot be filled to capacity when taking a full load of passengers and baggage.
▶ Always use actual measured weights if you are anywhere near the limits. Estimating the weight of baggage and passengers can result in significant errors.
▶ Dip the fuel tanks to get the correct fuel weight.
▶ Prior to the flight, check the movement of the center of gravity due to the planned fuel burn, as you may be out of center of gravity limits when you land.
▶ Check the POH for center of gravity or weight restrictions with respect to certain maneuvers (e.g. steep turns).
▶ Take care when calculating moment arms – arithmetic errors, and incorrect mixing of units are mistakes that can easily be made.
▶ If a passenger knows their weight, add an allowance for clothing and footware. (Better still put them on the scales.)
▶ Ensure the load (baggage) is properly secured. Accidents have resulted from loads moving during takeoff, resulting in the airplane going out of balance.

Never fly an overweight and/or out of balance airplane. Some pilots feel that a small amount doesn't matter, but it reflects an attitude that leads to unsafe behaviors, and a significant increase in risk.

Night Takeoff and Landing

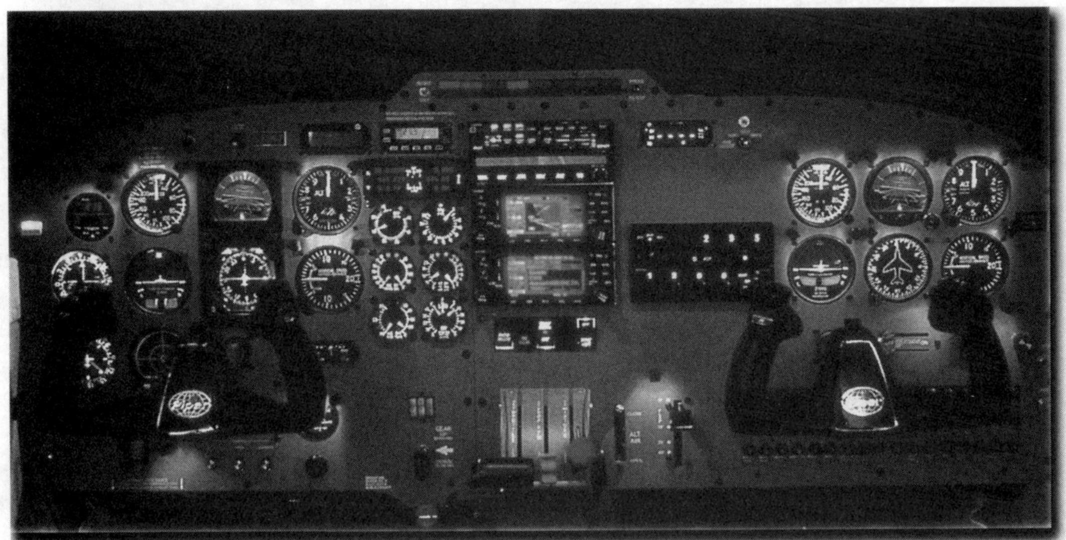

Courtesy of New Piper Aircraft, Inc.

Night takeoff and landing is subject to greater risk and is fraught with traps for the unwary or non-proficient pilot. Increased risks are due to:

- Reduced availability of visual cues.
- Vision that is less effective at night because the eyes function differently in darkness.
    - Night adaptation takes time. (Your eyes need about 30 minutes to adapt to maximum night efficiency after exposure to bright light.)
    - There is a need to view objects "off center" because of the different seeing mechanisms of the eye, compared to daylight viewing.
    - Fatigue, colds, vitamin deficiency, stimulants, smoking or medication can seriously impair your night vision.
    - Night vision acuity decreases with altitude above 5,000 feet without supplemental oxygen.
- A lack of color and contrast that results in you feeling further away and higher than you really are. There is therefore a tendency to fly approaches too fast and too low.
- Surrounding lights blurring airport boundaries, runways and taxiways.
- Well-lit roads being confused for runways.
- An increased risk of spatial and visual illusions.
- The greater ramifications of electrical failure.
- The likelihood that some landing strips along the intended route will not have lighting, and are therefore not available in an emergency.
- An increased risk of inadvertent entry into IMC.
- A lack of warning of suddenly reducing visibility.

Night takeoff and landing risks can be minimized as follows:

> ▸ Ensure you are well rested and in good physical condition ("I'M SAFE").
> ▸ Carefully plan the flight and consider your courses of action in case of possible deteriorating weather.
> ▸ Allow plenty of time for the pre-landing checks. Try to do them before the traffic pattern (you want to avoid distractions).
> ▸ Maintain the cockpit lights as dim as possible, and remember that night adaptation takes time (e.g. after using a bright torch).
> ▸ Maintain the recommended speeds in the pattern.
> ▸ Avoid long, low finals.
> ▸ Be familiar with obstacles and their location.
> ▸ Make sure the heading indicator is aligned with the magnetic compass, and set the heading bug to the active runway to assist in both finding the runway, and flying the pattern correctly.
> ▸ Make sure the altimeter is set correctly.
> ▸ Be aware of the various illusions that can occur at night.
> ▸ Make sure your flashlight is handy (and working).
> ▸ Always look slightly away from the object you wish to see.

Night flying can be a safe and enjoyable experience, provided due account is taken of the risks.

Avoiding Gear-Up Landings

Gear-up landings continue to occur with some regularity, with most due to lack of procedures, or lack of concentration.

Specific reasons for pilot induced gear-up landings include:

> ▸ Neglecting to extend the landing gear.
> ▸ Inadvertently retracting the landing gear.
> ▸ Activating the gear, but forgetting to check the position.
> ▸ Extending gear too late.
> ▸ A warning horn that is not operational (e.g. due to a pulled circuit breaker).
> ▸ Ignoring the warning horn.

Steps can, however, be taken to minimize the risk by the application of discipline and strict adherence to procedures.

> ▸ Try to have a set routine, and put the gear down at a standard point in relation to the inbound procedure being used, such as:
> > ▸ At the inbound point of a procedure turn.
> > ▸ At the final approach fix.
> > ▸ Abeam the landing runway.
> > ▸ At the start of descent from a pre-defined altitude, such as circuit altitude etc.

> ▸ Always have a back-up check in case the pattern changes.
> ▸ Be disciplined in carrying out pre-landing, gear-down checks.
> ▸ Be careful that passengers don't inadvertently bump the switch into the "Up" position (in some circumstances the locking mechanism can be rendered inoperative).
> ▸ Make certain you are familiar with the landing gear emergency extension procedure.

Furthermore, as noted by Thomas Turner in his article *Gear–up Landings*, check the indicating system as soon as possible after extension, but in addition have other ways of confirming that everything is normal. For example, listen for various noises that gear extension causes, and note any effect on the airplane attitude and airspeed. He also reminds pilots, not to depend on the warning horn, or other safety devices, to protect them from a gear-up landing.

Use of the above tactics will minimize the risk of a gear-up landing.

Flying Into Unfamiliar Unlicensed Bush Airstrips

Flying into unfamiliar unlicensed bush airstrips can be hazardous. They can be short, difficult to find, or have unknown obstacles. However, by taking suitable precautions the risks can be minimized (although pilots need to be cautious and not take unnecessary risks).

Possible Hazards

> - A very short or narrow runway.
> - A slope greater than 2%.
> - Trees may exist at either end and/or on the sides of the runway, acting as obstructions, or causing possible wind shear.
> - Possible hazardous surfaces that may not be visible from the air (e.g. rocks, soft ground, crossing tracks, holes, washouts, raised grass clumps, standing water, depressions).
> - Nearby power or telephone lines (difficult to see).
> - The presence of animals (cattle, sheep, wildlife).
> - A torn windsock or no windsock at all.
> - Non-standard runway markers, or none at all.
> - A lack of easily identifiable features making the airstrip difficult to locate.

Obtaining Airstrip Details

When planning to fly to a bush strip with which you are not familiar, or about which you do not have up to date information, it is necessary to contact the airstrip operator and obtain all the relevant airfield data. It is important to talk to someone who has aviation knowledge; otherwise you may not get the correct information. Never rely on a statement such as "The strip is in good shape and you should have no difficulty". Obtain specific details of the following:

> - Location
> - Elevation
> - Runway direction, usable length, width and slope.
> - Runway surface type (sealed, broken sealed, soil, sand, gravel, clay, grass, naturally hard, naturally soft).
> - Condition of the runway surface (cracked, sandy, hard, soft, muddy, hardened mud, heavily grassed, lightly grassed). If the surface is grass, when was it last cut and how high is the grass? As a rule of thumb, it should be no more than 30% of the diameter of the main wheels of the airplane.
> - Whether it has recently rained or not (the ground may look dry but could be unacceptably soft underneath). Is the surface dry, moist or wet?
> - Usability of the runway. Does it have machinery, stock or wildlife on it? Is it still being used as a runway? Does it have any holes, drains, mounds, rock, wash-aways etc?
> - Location and height of any trees, radio masts, power lines, telephone lines, buildings or any other obstructions on the approach and takeoff areas.
> - Direction of the prevailing winds, and any local effects (e.g. wind shear caused by trees or other obstructions).
> - Wind sock location, color and condition. If there is no windsock, what help can be provided to determine the wind direction and speed.
> - Types of airplanes that have recently used the strip.
> - Landmarks and features to allow identification from the air.
> - Nearby terrain (e.g. flat, hills, structures, power lines).
> - Local procedures.

> ▸ The nature and condition of the movement and run up areas.
> ▸ The means of identifying the boundaries of the movement area.
> ▸ Any restriction as to the use of the field.

Some Checks

The first step is to check the airplane takeoff and landing performance relative to the identified conditions and to allow an adequate margin of safety. Make sure the landing area really is long enough. Then review your own proficiency in relation to short field landing and takeoff (if that is going to be required). In addition:

> ▸ Ensure that you positively identify the field, otherwise head for your alternate.
> ▸ Carry out a precautionary approach (see Section 14).
> ▸ Establish the wind direction and speed.
> ▸ Carry out the correct landing for the conditions (e.g. short field landing if the field is short, or a soft field landing if the field is soft).
> ▸ Be prepared to go round if your approach is not correct or you are not going to land at your predetermined touch down zone.
> ▸ After touch down, apply firm braking with the control column back, and watch directional control.

If in doubt about the quality of the airstrip information or the airstrip condition, then err on the side of caution and land elsewhere.

Establishing Wind Direction and Speed

If there is a regulation windsock, the wind velocity can be estimated from the angle the sock makes with the vertical support post. The windsock can also tell you whether the wind is gusting or not, and the degree of gusting.

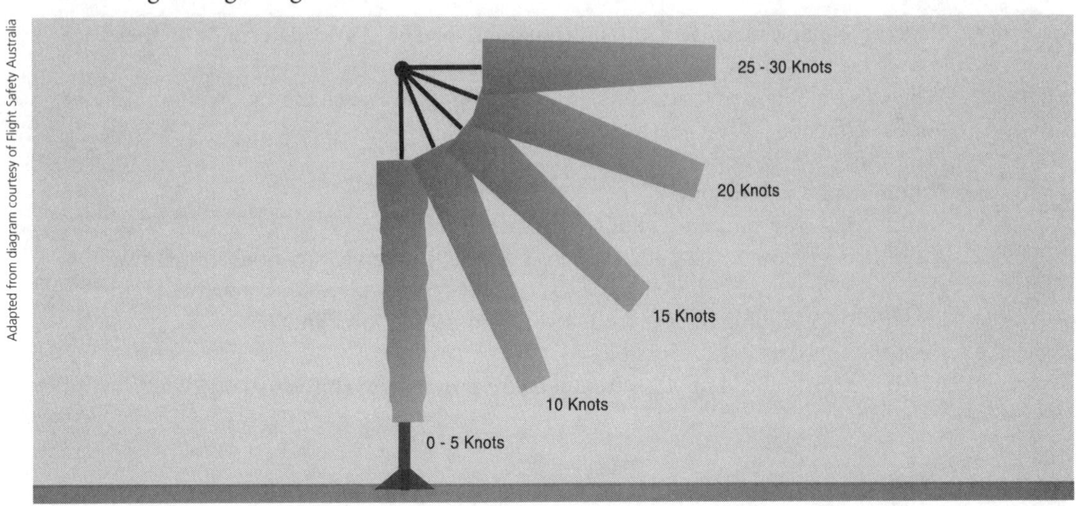

If there is no windsock, you can use other observations as noted in the table below.

Description	Wind Speed	Observation
Calm	0 - 1 kts	Smoke rises vertically.
Light air	1 - 3 kts	Wind direction shown by smoke drift.
Light breeze	4 - 6 kts	Wind felt on face. Leaves rustle.
Gentle breeze	7 - 10 kts	Wind extends light flags. Leaves and small twigs are in constant motion.
Moderate breeze	11 - 16 kts	Raises dust and loose papers. Small branches move.
Fresh breeze	17 - 21 kts	Small trees in leaf sway. Crested wavelets on water.
Strong breeze	22 - 27 kts	Large branches in motion.

Bird Hazards

Bird strikes can potentially inflict considerable damage. Generally birds have enough time to get out of the way of an airplane flying at 80 knots but above 100 knots the incidence of strikes increases. Fortunately, most bird strikes do not result in significant damage although there are the exceptions. It is therefore wise to take avoidance action whenever there is the potential for bird strikes.

Avoidance Strategies

Most bird strikes occur up to 1,000 feet, although they have been known to occur at middle and higher altitudes. In areas subject to birds, the following actions will help reduce the possibility of a bird strike:

- Check the relevant airport publications and NOTAM's to determine if there are known bird hazards.
- When aware of bird activity, consider changing runways, changing the flight time or not flying at all.
- Plan to fly as high as possible.
- Maintain a good look out.
- Avoid flying over bird and wildlife sanctuaries.
- Avoid flying below 1,500 feet above ground level in the vicinity of abattoirs, swamps, garbage dumps, rivers, shorelines, lakes, estuaries or any place likely to attract birds.
- Fly with your landing lights, strobe lights and navigation lights on.
- Keep the speed as low as possible.
- If you see birds ahead of you, attempt to fly above them (birds usually dive to avoid a collision).

Guarding Against Illusions

9

Many of the decisions we make in the cockpit result from what we see or what we sense. However, *what we see and what we sense may not always be correct*, as the result of various forms of illusions. The most well known illusion occurs when a non-instrument proficient pilot inadvertently flies into cloud and becomes disoriented, due to vision being unable to counter the signals from the inner ear. You can literally be flying upside down and not know it! In addition, there are numerous visual illusions, ranging from the effect of rain on the windscreen, to illusions caused by sloping runways, most of which occur during approach and landing. Pilots need to be familiar with these so as to be able to counter their potentially hazardous effects.

Spatial Disorientation

Spatial disorientation can result, when there are no visual clues to help counter the various sensory mechanisms, particularly the organ of balance that senses linear and angular acceleration. Lack of visual clues can occur when flying in cloud, in poor visibility, and at night, without reference to the airplane instruments. Disorientation can result when the pilot fails to correctly sense the position, attitude and motion of the airplane as the result of the illusion. Spatial disorientation is a well-recognized cause of aviation accidents to which, even very experienced pilots, are susceptible.

Types of illusions that can lead to spatial disorientation are:

The "Leans"

The "leans" results when a pilot abruptly recovers from a slowly entered turn, and incorrectly senses that the airplane is banked in the opposite direction. There will be a tendency to adopt the original banked attitude, or lean in the perceived vertical plane until the illusion subsides.

Graveyard Spiral or Somatogyral Illusion

In a steady rate coordinated turn, there is a false sensation of straight and level flight, and hence, a gradual entry into a turn or spiral, can go undetected until a dangerous degree of rotation has resulted. An extreme situation is the graveyard spiral, which results when the pilot observes a loss in altitude with the illusion of level descent, and pulls back on the controls, only to aggravate the spiral.

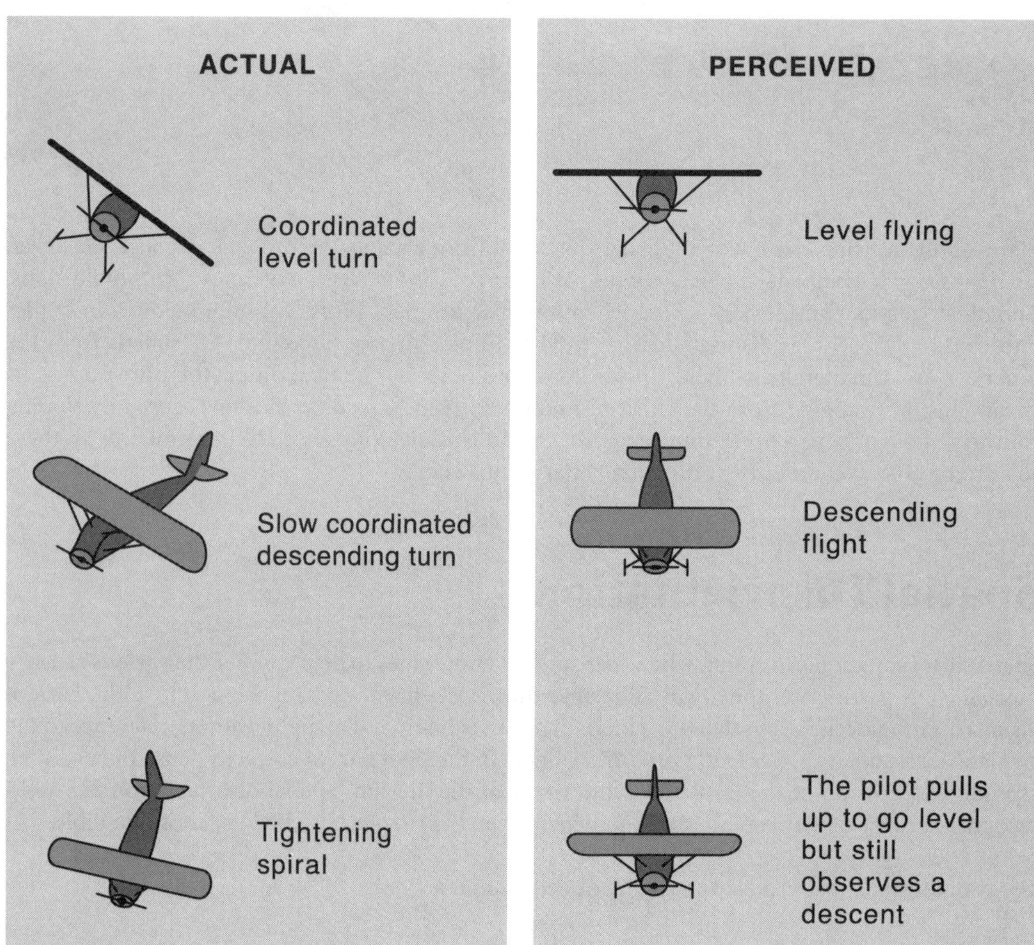

Coriolis Illusion

An abrupt head movement in an established constant rate turn, can create the illusion of rotation or movement on an entirely different axis, with the tendency to correct inappropriately. This illusion can be very dangerous because of the overwhelming disorientation power.

Somatogravic or "False Climb" Illusion

In the absence of visual clues, acceleration of an airplane and tilt of the head can cause the balance organs, and the brain, to conclude that the airplane is climbing steeper than it actually is. Unless countered by reference to instruments, the pilot will instinctively pitch forward. Situations in which this illusion can occur are during:

> ▸ Takeoff. (This is particularly dangerous when taking off from an airport at night where there is no surrounding light, or where there is low cloud - the pilot simply flies into the ground at high speed.)
> ▸ An overshoot (missed approach).
> ▸ A climb from visual flight rules, into instrument rule conditions.

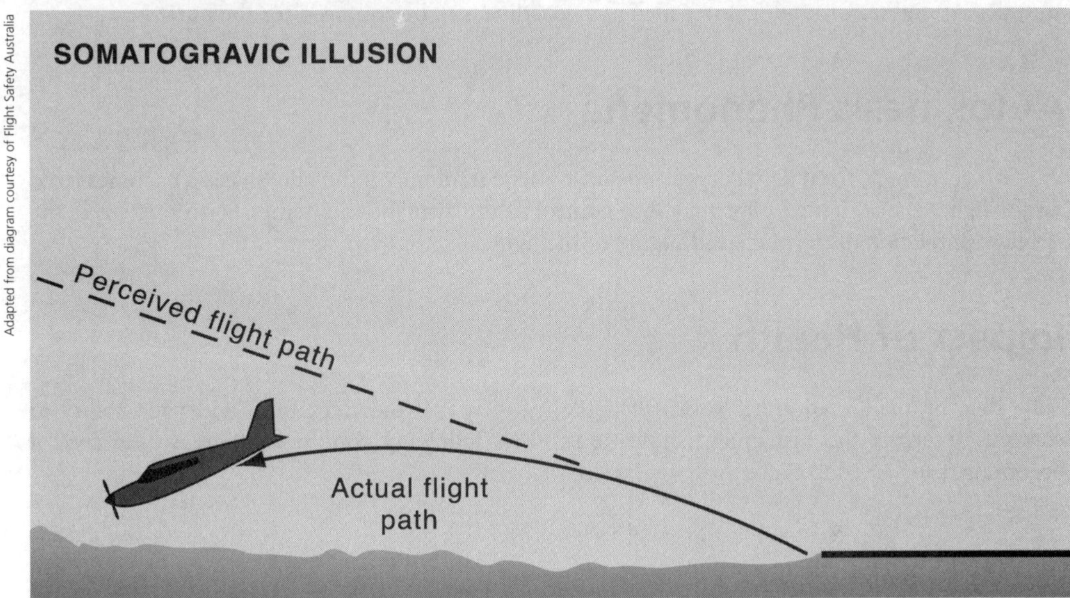

Adapted from diagram courtesy of Flight Safety Australia

"Reverse" Somatogravic Illusion

Deceleration can cause the false sensation that the airplane is descending, with the pilot reacting by raising the nose, and possibly further heightening the illusion. An aerodynamic stall is then a possibility.

Inversion Illusion

An abrupt change from a climb to straight and level flight can create the illusion of tumbling backwards, with the resulting tendency to push the nose abruptly down, possibly making the illusion worse.

Elevator Illusion

An abrupt upward vertical acceleration (e.g. due to an updraft) can create the illusion of being in a climb. A disoriented pilot will push the airplane into a nose-down attitude. A downdraft gives the opposite effect.

False Horizon

A lack of a defined external horizon, caused by sloping cloud formations, a line of lights along a highway, mountains, or high ground, can create an illusion of not being aligned correctly with the horizon, with a disoriented pilot placing the airplane in a banked (and possibly dangerous) attitude. At night, a ground feature, such as a coastline, can be confused for the horizon.

Autokinesis Phenomena

In this case, a single fixed light source appears to move randomly if the pilot's vision becomes fixed on the light. A disoriented pilot may lose control of the airplane in attempting to maneuver the airplane in relation to the perceived nature of the light.

Impact of Health

Your state of health can affect your susceptibility to spatial disorientation, and your ability to correctly interpret the instrument indications. The following conditions increase the risk of becoming disoriented in flight:

- Fatigue.
- Alcohol.
- Certain medications.
- Low blood sugar.
- Dehydration.
- Middle, and inner ear medical disorder.
- Stress.
- Minor illnesses.
- Hypoxia.

Avoiding Spatial Disorientation

Pilots should know the circumstances, and conditions, that can lead to spatial disorientation and be vigilant for their possible occurrence. Be aware that you cannot always recognize that an illusion is occurring unless you are regularly cross checking the instruments.

If you are not instrument qualified, then do not attempt visual flight in conditions that can lead to loss of external visual reference. Avoid unnecessary, abrupt head or airplane movements, and do not fly when affected by conditions that aggravate sensory illusions.

If you find yourself flying in conditions favorable to illusions that can result in spatial disorientation, transit onto instruments quickly. Ensure that you rely on the instrument indications, unless the natural horizon, or surface reference is clearly visible. Maintain a regular crosscheck of the instruments and do not attempt to mix visual and instrument flight until the visual clues are unmistakable.

Note that you can lose orientation in less than 20 seconds if you are in cloud, and not on instruments. Lack of instrument proficiency, under these circumstances, can be deadly.

Approach and Landing Errors

There are a number of visual illusions that can lead to false, and conflicting, perceptions during the approach and landing phases of flight. They are therefore potentially dangerous if not understood and countered, particularly those that occur at night. The following situations and conditions result in illusions that can lead to approach and landing errors.

Shallow Fog

While the runway may initially be visible, the visual reference decreases as you enter the fog. This may cause an illusion that the airplane has pitched upward, leading the pilot to make an inappropriate nose-down correction.

Runway Lighting Intensity Illusion

On clear nights, runway lights may appear closer than they actually are, particularly if there is no light in the surrounding area. This can give the illusion of being too high.

Sloping Runway Illusion

An upsloping runway can create the illusion that the airplane is at a higher altitude than it actually is, resulting in a tendency to fly a lower approach. A downsloping runway has the opposite effect.

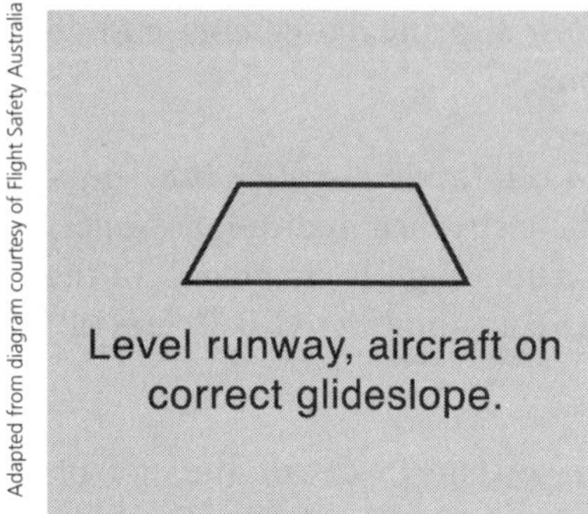

Level runway, aircraft on correct glideslope.

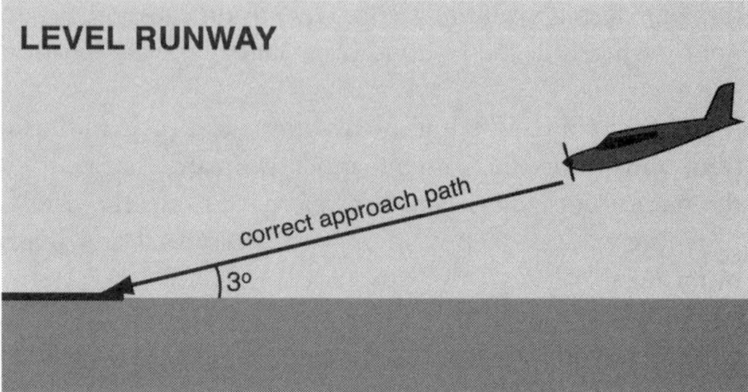

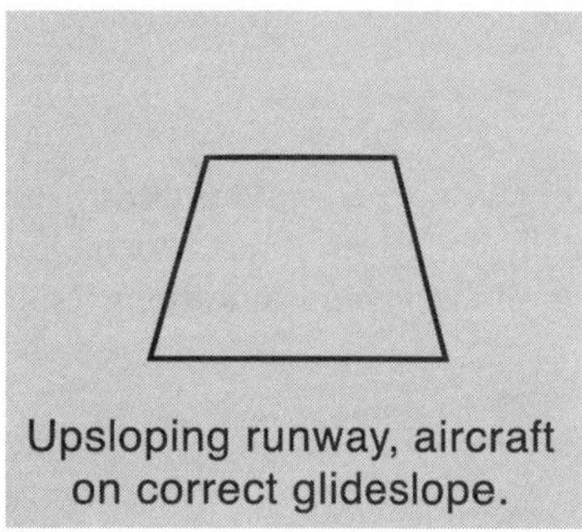

Upsloping runway, aircraft on correct glideslope.

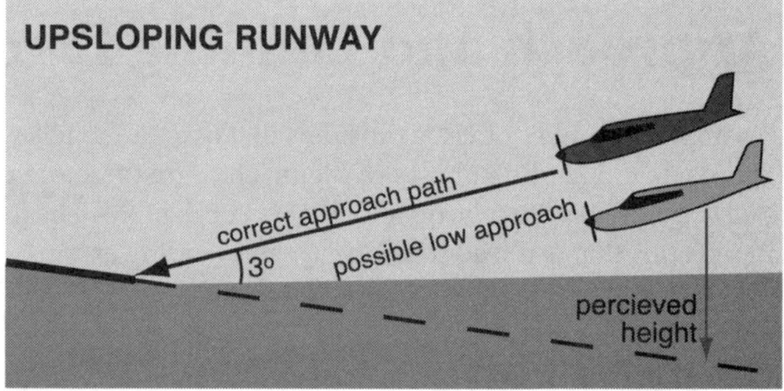

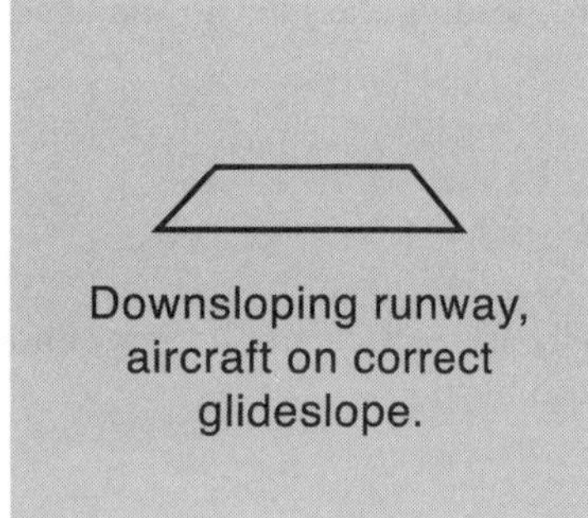

Downsloping runway, aircraft on correct glideslope.

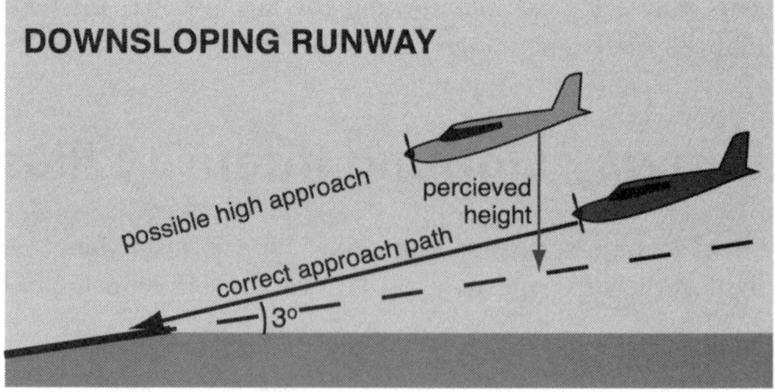

Sloping Approach Terrain Illusion

The illusion due to sloping approach terrain is similar to that resulting from sloping runways. The greatest hazard is with upward sloping terrain, where again, the tendency is to move below the required approach path, as there is a sensation that you are too high.

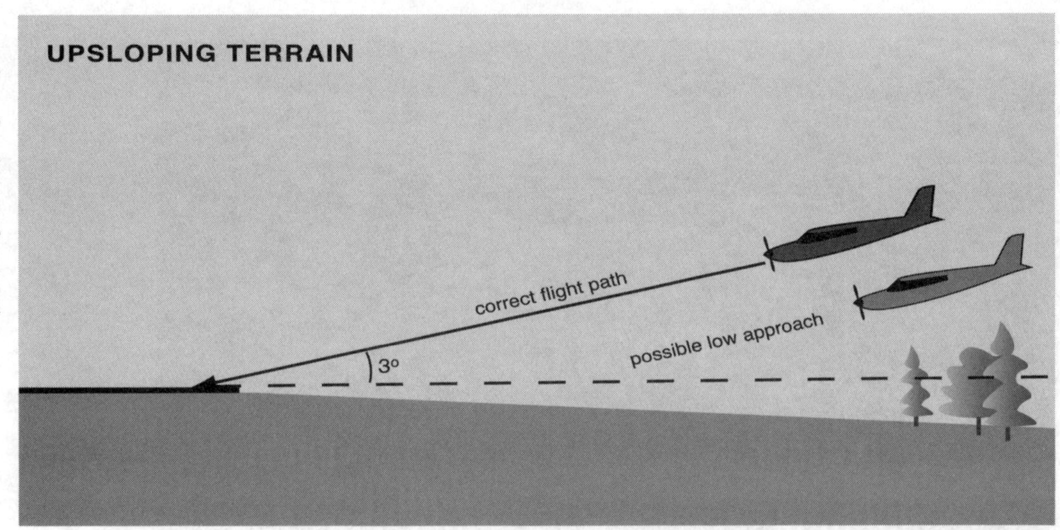

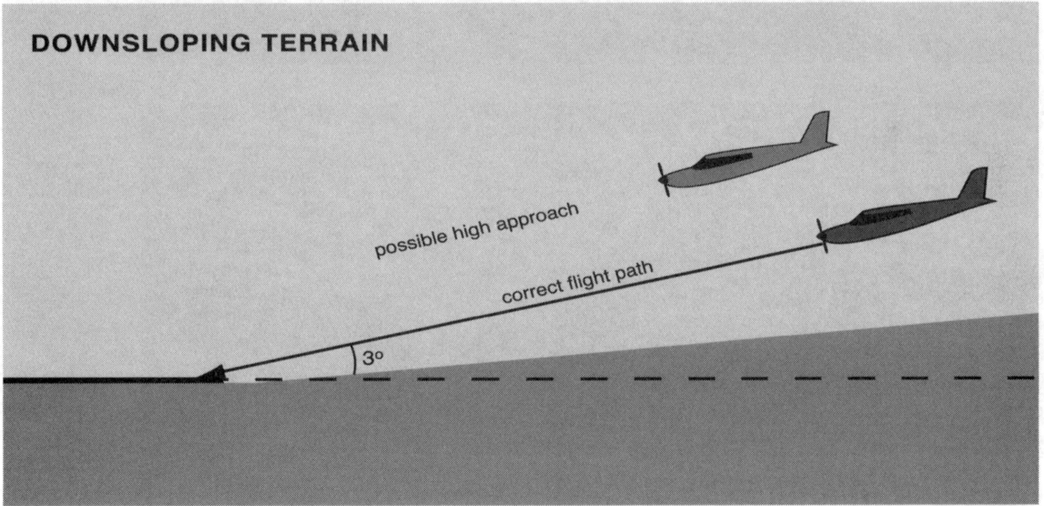

Runway Width and Length Illusion

Narrower runways, relative to what you are familiar with, give the illusion that the airplane is higher than it actually is, resulting in a lower approach path to the runway. A longer runway gives the same impression, and again, is potentially hazardous. A wider or shorter runway can have the opposite effect, resulting in too high a round-out, and landing hard, or overshooting.

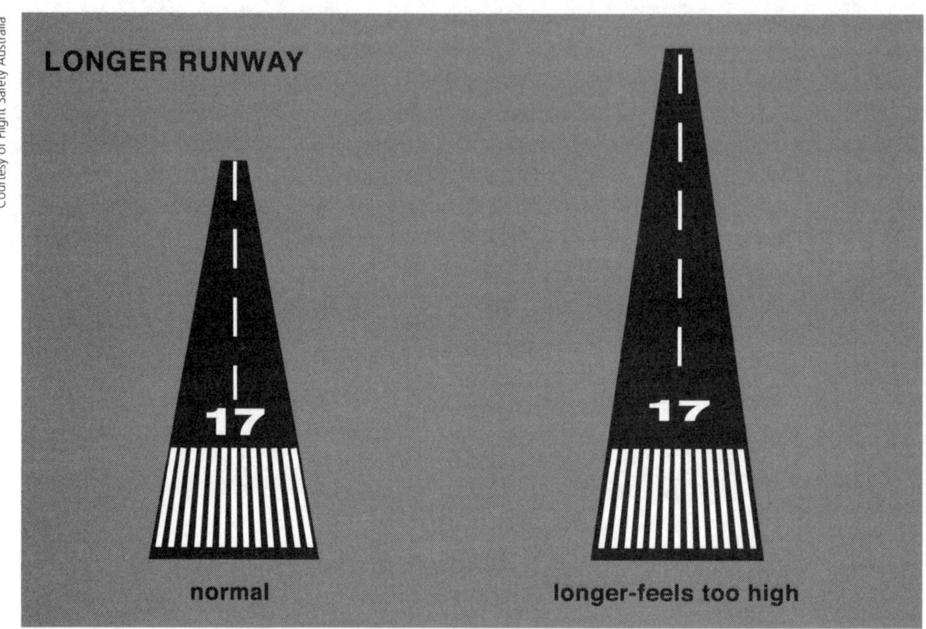

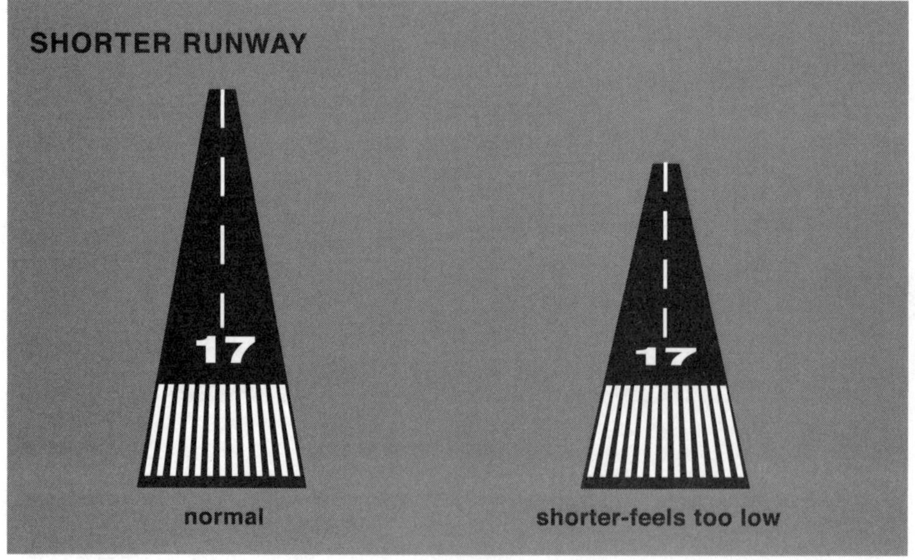

Rain Effects

Rain falling on the windshield and on approaching terrain, can cause several visual illusions, as the result of refraction and diffusion, depending on the circumstances.

> ▸ Terrain contours, or lights, may appear lower than they actually are, due to refraction, which could lead to a premature descent, or inaccurate assessment of the terrain clearance.
>
> ▸ Lights may appear to be less intense, due to diffusion (halo effect), and therefore further away than they really are.
>
> ▸ Under other circumstances, diffusion may cause approach lights to appear larger, giving the illusion you are closer than you really are, leading to a premature descent.

Black Hole Effect

A pilot flying over featureless terrain at night, with the intensity of the runway lights the only visual clue, may fly a lower than normal approach because of limited depth perception. This is generally known as the "black hole" effect. Some scenarios which make the effect more pronounced include:

> ▸ An airport that is on the near side of a brightly lit city, with few, or no, terrain features or lights between you and the airfield.
>
> ▸ An airport that is on the coast, or in very sparsely settled terrain, such as deserts or wilderness areas.
>
> ▸ Nights with extremely clear air and excellent visibility.
>
> ▸ An airport that is at a lower elevation and/or at a different slope than the surrounding terrain.
>
> ▸ At an airfield where the runway lighting is poor, and other landing aids are either unavailable or unserviceable.

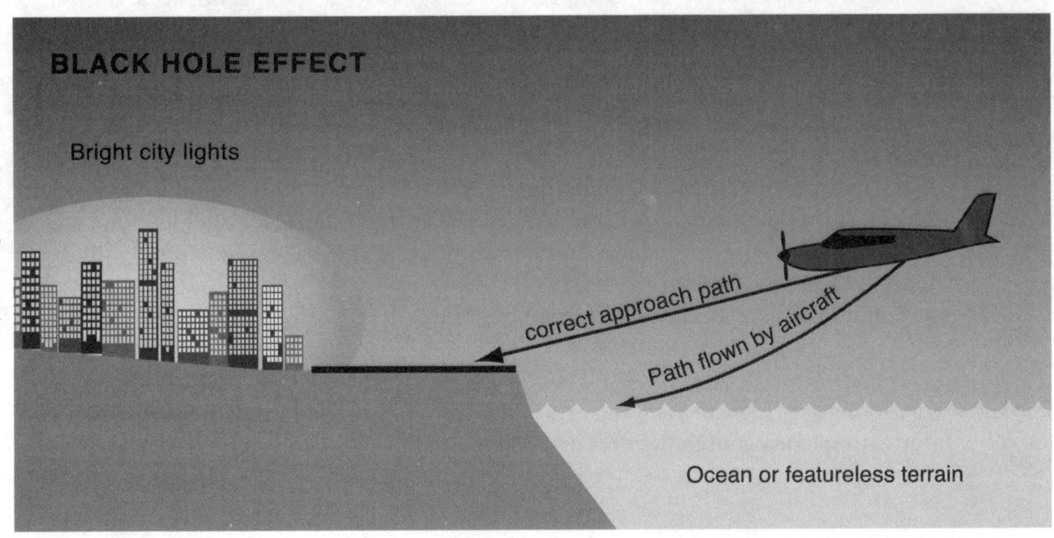

Ground Light Confusion

Visually locating an airport and its runway can be difficult because of the confusion created by other lights. For example, road lights can be mistaken for runways, and stars can be confused for ground lights (and vice versa).

Haze Effect

Flying in haze can create the illusion that the runway is further away than it really is.

Countering Illusions that Can Lead to Approach and Landing Errors

It is unlikely that pilots will be able to recognize every illusion, nor the extent of that illusion. However, there are measures that can be taken to avoid being caught.

> - Be aware of the circumstances that contribute to the various illusions, and fly with extra attention and caution.
> - Check the airport and surrounding terrain for features that can lead to illusions.
> - Fly a standard circuit so that height can be checked against known positions in the pattern.
> - Avoid long, straight-in approaches at night, without guidance from a glide slope indicator.
> - Scan the instruments regularly so as to remain aware of the airplane's airspeed, altitude and sink rate.
> - Be careful in the transition from an instrument approach to a visual approach. Be prepared to take corrective action if necessary (e.g. a go-around or a missed approach).
> - Refer to the attitude instruments frequently when flying at night, or in reduced visibility.
> - Obtain experience at airports/conditions that result in illusions, particularly at night.
> - Use your instruments and/or glide slope lighting to monitor the glide slope.
> - At night, maintain sufficient altitude until the airport, and associated lighting, are clearly visible and identifiable, and then do not descend below the circuit altitude until in the circuit pattern.
> - Maintain sufficient altitude and terrain clearance.

The key to countering approach and landing error illusions is to be aware when they can occur, and use other information to assist in flying the correct flight path.

Mid-Air Collision Avoidance

10

While mid-air collisions are not common, they do occur, and near misses happen more often than we might imagine. It is the responsibility of the pilot to "see and avoid" whenever weather permits, regardless of whether the operation is conducted under IFR or VFR. However, the "see and avoid" principle has a number of serious limitations, and therefore, collision avoidance requires constant vigilance, an effective scan pattern and use of all other avoidance means available to a pilot. *No pilot, irrespective of the level of experience, is immune to the risk of collision.*

Limitations of "See and Avoid"

A basic understanding of the "see and avoid" limitations is necessary for effective collision avoidance.

Limitations of the Eye

While a marvelous instrument, the eye has numerous deficiencies with regard to air traffic detection.

- The eye is vulnerable to the vagaries of the mind. For example, if you are daydreaming and simply stare into space, you are unlikely to see approaching traffic.
- It takes up to 2 seconds or more to refocus the eye when looking up from the panel and trying to see an object a mile or more away.
- When the head is in motion, vision is blurred, and the mind will not register potential targets.
- Another focusing problem results in "empty-field myopia" – the eye has difficulty focusing on empty space and tends to focus at a distance of about 20 feet.
- The eye has a narrow field of effective vision, limited to about 10° to 15°. If an object does not fall within this cone, you may not see it, as visual acuity drops by 90%.
- Motion or contrast is needed to attract the eye's attention. Unfortunately, airplanes are most difficult to see when they are on a collision path, because the relative position remains constant. (See diagram opposite.)
- The human visual system also suffers from optical illusions. For example, an airplane flying towards you at a slightly lower altitude, will appear to be above you, and then appear to descend, as it comes closer.
- An airplane against a cluttered background (e.g. buildings, the ground) will blend into the background until it is quite close.
- For the brain to accept what we see, we need to receive cues from both eyes. An airplane "seen" by one eye, but hidden from the view of the other eye, (e.g. by the windshield post or other obstruction) results in an image that is not always accepted.
- Fatigue, sleep deprivation, emotional stress, age, smoking, alcohol and medication can all affect a pilot's vision.
- Glare, haze, windscreen cleanliness/distortion all limit, restrict or distort vision.

Other Limitations

The low frontal profile of most light airplanes, and their relatively small size, makes detection extremely difficult, even if the position is known.

- Even at the closure rate of most light airplanes, there is only a limited amount of time between visual detection and potential collision.
- The nature of cockpit and airplane design, results in a rather restricted field of vision.
- While a radio aids in collision avoidance, not all airplanes are radio equipped.

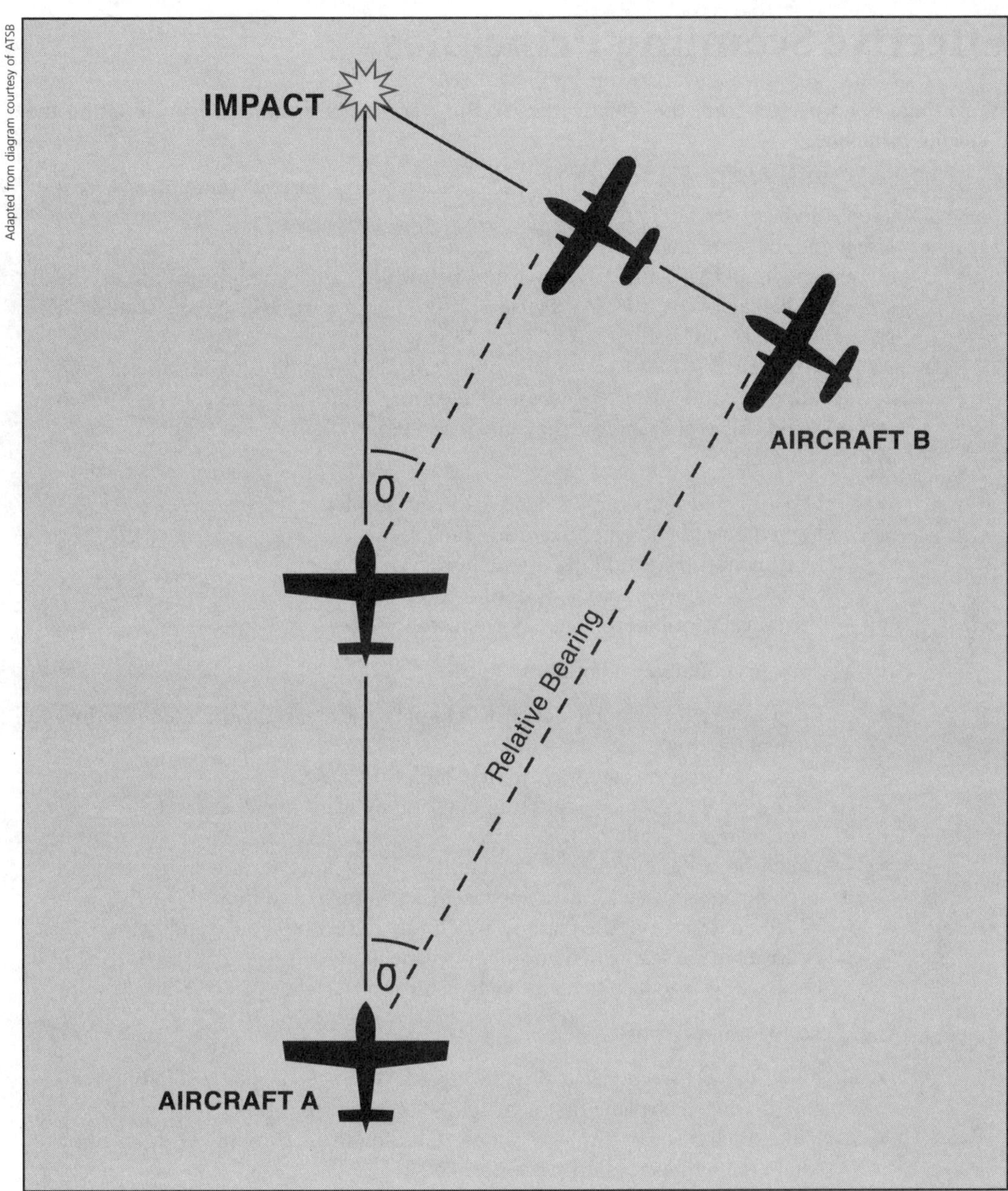

As a consequence of these limitations, it is necessary to adopt an effective scanning technique and compensate for the inadequacies of visual detection by other means.

Effective Scanning Principles

While there is no perfect scan, the effectiveness of the scan can be maximized by adopting the following principles:

- Don't overestimate your visual abilities, and understand the eye's limitations. (Most people, in fact overestimate their capability.)
- Use a definite scanning pattern that concentrates on the most critical areas. Scan a 60° area right and left, and 10° above and below your flight path, breaking up the scan into blocks of 10° to 15°. Pause briefly (1 to 2 seconds) as you scan each segment. Two established techniques are shown.
- As you begin each scan, focus initially on the most distant object you can see such as a cloud, a building or other distant visible ground feature. This will take 3 to 5 seconds, and will avoid saccadic movement and empty-field myopia.
- When scanning:

    - Look for stationary targets relative to your airplane.
    - Keep the head still when scanning each sector.
    - Don't concentrate too long on the horizon.
    - Don't stay focused on one object for too long.
    - Periodically look beyond the normal scan limits.

- Always look for potentially conflicting traffic.

    - Keep a sharp lookout, particularly in high traffic areas. Ensure adequate time is spent looking out of the window.
    - Get your co-pilot/passengers to help look for traffic.
    - During final approach, avoid tunnel vision by scanning all around.
    - Scan for traffic while on the radio.
    - Watch both the circuit and outside the circuit for entering traffic (e.g. an airplane on a low straight-in approach as you turn onto final).
    - When on descent in the pattern, watch for traffic both above and below.
    - Don't relax the scan when another airplane is sighted.
    - Use passengers to help maintain sight of an airplane while you continue to scan.

- Compensate for blind spots.

    - Avoid the well-recognized situation of a low-winged airplane colliding with a high-winged airplane that is on a lower flight path on final.
    - Make small clearing turns every now and then while climbing or descending (there is a blind spot under the nose).
    - Make sure the area is clear before any turn. High wing airplanes should pick up a wing before turning to check for traffic.

When the external scan has been completed, scan the instruments before repeating the procedure. Develop an efficient time-sharing regime. Remember, you are looking for a small target that gets bigger.

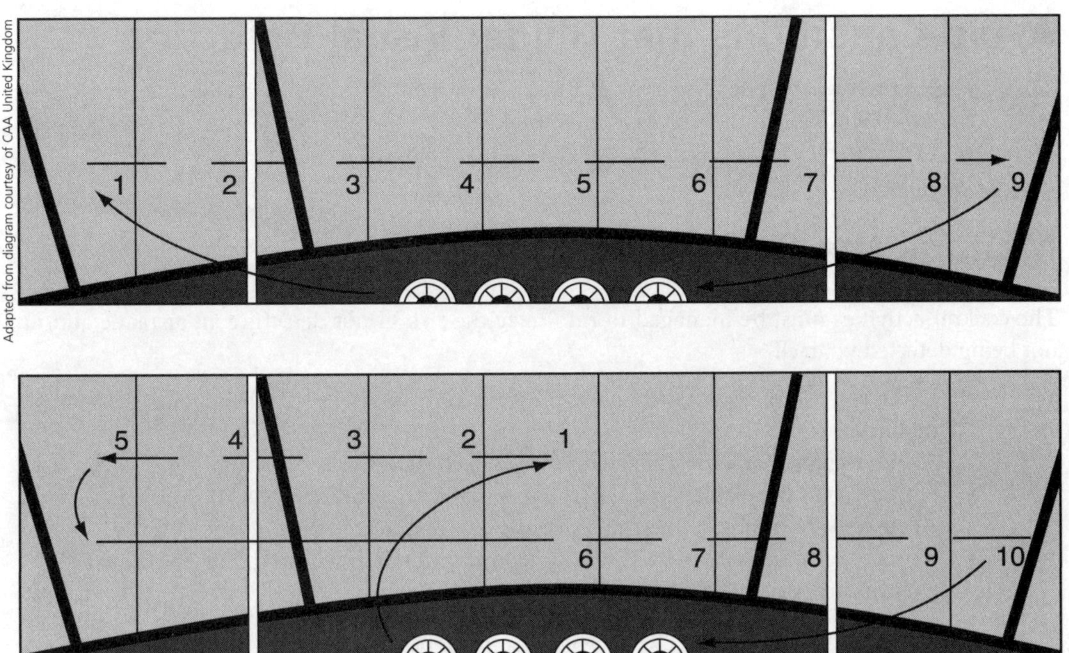

Additional Collision Avoidance Measures

Bearing in mind that there is no foolproof visual procedure, a pilot needs to utilize every additional avoidance measure that is at his or her disposal, in order to minimize the risk of a collision.

Be Aware of High Risk Environments

While mid-air collisions have occurred in many situations, the majority of collisions occur:

- During daylight with good VMC.
- At low altitudes.
- Near airfields, and in the pattern.
- While overtaking.

Being extra vigilant under these circumstances is essential to minimizing the risk of a collision. During preflight planning, check for high traffic density areas, special use airspace and military areas. Try to avoid crowded airspace, and remember that some traffic may not be where they are supposed to be.

Avoid Conditions that Hinder Visual Detection

- Ensure you are in a good state of health, both mentally and physically.
- Avoid smoking and control caffeine intake.
- Maintain a clean windscreen.

Manage the Cockpit Activities

The cockpit activities must be managed to minimize the risk of not detecting an airplane and/or not being detected yourself.

- Don't get mesmerized with head-down cockpit work. Plan ahead and have an organized cockpit.
- Avoid doing paperwork during climb and descent.
- Stay ahead of the airplane and maintain situational awareness.
- Avoid any absentmindedness and avoid being distracted.
- Reduce or eliminate non-essential communication while flying in high-risk environments.
- Avoid communication breakdown.

Talk and Listen to the Radio

The radio provides a very effective way to reduce the risk of a mid-air collision, if used to its full potential.

- Listen to the controller, and other pilot's radio calls and develop a mental picture of the position of other airplanes in your vicinity. Query any clearance or instruction if there is uncertainty.
- Give meaningful and accurate position reports (note whereabouts and intentions).
- Use ATC services if available.
- Make required radio broadcasts in the pattern so as to give other pilots a place to look for your airplane. Also, remember that banking airplanes are easier to see.
- Be aware that there are airplanes without radios or simply not making broadcasts.
- If uncertain about an airplane's position, ask for its location. If the airplane is still not sighted, take positive action to avoid a potential collision.

Utilize Anti-Collision Equipment

- Have the transponder on and set to altitude.
- Turn on the landing lights when within 10 nautical miles of a busy aerodrome (particularly in hazy weather).
- Make sure the strobe lights are turned on (except in cloud).

Follow Established Procedures and Regulations

Standard operating procedures and regulations are designed to ensure an orderly flow of traffic so as to reduce the probability of a collision, as well as allowing controllers and pilots to anticipate where traffic might be. It is particularly important to adhere to procedures no matter where you are, or whether or not there is any known traffic. Some specific reminders are:

> ▸ Fly the correct en-route altitudes for the direction of flight (both VFR and IFR).
> ▸ Always follow published departure, approach and circuit entry procedures.
> ▸ Be at pattern altitude before entering the pattern.
> ▸ Fly the correct circuit pattern.

However, always be on the lookout for pilots deviating from correct practices (e.g. wide circuits, incorrect circuit entry, non use of the radio).

Other Measures

> ▸ When flying IFR, be aware of when you need to use the "see and avoid" principle.
> ▸ Don't relax attentiveness and vigilance when in controlled airspace, or when the weather is good.
> ▸ Slow the airplane down in the vicinity of the traffic pattern.
> ▸ Avoid flying (en-route) near airports without maintaining adequate lateral and altitude margins.
> ▸ Avoid landing into the sun (if possible).
> ▸ Don't assume anything (e.g. that you are the only airplane in the vicinity).
> ▸ Check for traffic on final before lining up, even if you have received a clearance.
> ▸ Make gentle S-turns while climbing or descending to check for traffic.
> ▸ Look both right and left before making a turn.
> ▸ Don't deliberately violate the rules.

When weather conditions permit, "see and avoid" remains the final backstop in avoiding mid-air collisions. Being familiar with the full range of anti-collision strategies and tactics will result in the risk of a mid-air collision being minimized.

Avoiding Engine Failure

11

Engine failure due to mechanical malfunction remains a relatively rare event, although it does occur. Reciprocating engine mechanical reliability depends upon proper maintenance, and *pilots operating the engine within the specified parameters and limitations*. Furthermore, engines usually provide early warning of a possible problem, and pilot awareness of abnormal indications can readily avoid a serious situation from developing. A pilot, therefore, can do a great deal to minimize the probability of an engine failure by performing some engine inspection, making sure that they operate the engine in accordance with the manufacturer's recommendations, and by monitoring the engine's condition. In addition, not all engines are identical, and pilots, therefore, need to have an appreciation of the differences in terms of these aspects of engine management.

Pilot Preflight Engine Inspection

Private pilots generally do little in the way of engine inspection other than checking the oil, draining the fuel filter, checking the alternator belt, and a cursory look for bird nests (without removing the cowling). They rely on the regular mandatory maintenance inspections to detect any anomalies. However, there are some simple checks that pilots can easily undertake, as outlined by John Schwaner in his AVweb article *How to Monitor Your Engine's Condition*.

> ▸ Look for signs of leaking oil (including the propeller hub and underneath the airplane).
> ▸ Examine the exhaust pipe by touching the inside edge with your fingertip (CAUTION: only if the exhaust pipe is cool). If the mixture setting and oil consumption are normal, then your finger should be clean, or possibly have a slight tan ash deposit. If your finger has dry black soot on it, then the engine is operating with a rich fuel/air mixture. If your finger has oily black soot, then the engine is burning too much oil.
> ▸ Smell inside the engine compartment for fuel smells, which could indicate a fuel leak.
> ▸ Check the color of the oil on the dipstick. If it looks like black lacquer, then the piston rings are leaking combustion gas into the oil.

In addition, check for possible exhaust system leaks by examining the cowling, aft of the engine, for grayish or brownish trails.

Some airplanes have sections of the cowling that are hinged. If this is the case, it is good practice to inspect those parts of the engine that are accessible, checking to see that:

- ▶ Everything is secure.
- ▶ There are no leaks (e.g. oil leaks, fuel stains, exhaust stains).
- ▶ There is nothing broken.
- ▶ There are no "foreign" objects (e.g. bird nest, rag).
- ▶ No wires have frayed.

Avoiding Pilot Induced Engine Damage

Pilots can induce engine damage if they operate the engine incorrectly or outside the normal operating envelope. Pilots should, therefore, be cognizant of the causes of engine damage. For example, thermal cycles, rather than hours of operation, is the major contributor to engine wear, and hence, unnecessary engine starting should be avoided. The table opposite gives some typical, inappropriate operating practices and the potential damage that may result.

Some Causes of Induced Engine Damage

Inappropriate Operating Procedure	Result	Potential Damage
▶ Unnecessary engine starts.	▶ Thermal cycles.	▶ Reduced engine life.
▶ Running at idle immediately after cold start.	▶ Inadequate splash lubrication.	▶ Unnecessary wear.
▶ Prolonged operation below 1000 rpm.	▶ Inadequate engine temperature.	▶ Spark plug fouling. ▶ Deposits on exhaust valve stems resulting in accelerated exhaust valve guide wear.
▶ Excessively lean mixture under high power conditions.	▶ Possible detonation and pre-ignition.	▶ Engine damage. (Detonation and pre-ignition can lead to dished piston heads, collapsed valve heads, broken rings, eroded portions of valves. Detonation cannot be heard in the cockpit.)
▶ Excessively rich mixture.	▶ Lower spark plug temperature.	▶ Spark plug fouling.
▶ Exceeding the optimal CHT by running too hot over a prolonged period (either on the ground or in the air).	▶ Excessive CHT.	▶ Cylinder head cracking.

Avoiding the above conditions, and operating the engine strictly in accordance with the POH and the engine manufacturer's recommendations, can maximize engine life and reliability. The following general advice on the care of engines is applicable to most airplanes.

> - Make certain the correct grades of fuel and oil are used, and that both are clean.
> - Idle at the appropriate speed specified in the POH to ensure sufficient engine lubrication occurs.
> - On start up, if the oil pressure is not in the "green" within 60 seconds, shut down and investigate.
> - Do not allow the engine to develop significant power before the oil has warmed sufficiently. Warm a cold engine for about 5 minutes before conducting the ground run-up.
> - The engine should not be run at high power setting for prolonged periods during run-up to avoid overheating and hot spots.
> - Keep ground operations to a minimum, consistent with warming up the engine, especially in hot weather, to avoid over-heating and hot spots.
> - Increase the throttle smoothly and slowly.
> - Once obstacles are cleared, climb at cruise climb speeds (improves cooling and allows for better vision over the nose).
> - Ensure the engine is leaned when required, as per the POH.
> - Avoid unacceptably high cylinder head temperatures. The CHT and oil temperature can be reduced (if too high) by:
>   - Opening the cowl flaps.
>   - Increasing airspeed (lowering the nose).
>   - Reducing power.
> - Avoid rapid cooling by keeping descent rates below 1,000 feet per minute, with sufficient power being maintained to keep the engine gauges in the green.

Avoiding Spark Plug Fouling

Spark plug fouling can be minimized, or even eliminated, by maintaining proper operating temperatures and the correct mixture settings.

> - Avoid running the engine below 1,000-1,200 rpm during ground operation to avoid spark plug fouling. Fouling detected during the run-up check, can generally be removed by running the engine with a lean mixture at 2,000 rpm. However, note that prolonged running can result in excessive CHT and detonation.
> - Do not let the engine idle for long periods.
> - Maintain CHT in the green.
> - Use the recommended leaning procedures as per the POH.
> - Avoid long, low-power descents.
> - Operate with at least a small amount of power on landing approaches.

Inactive Engines

Engines not flown frequently deteriorate far more rapidly than those flown consistently. Products of combustion include water, which accumulates in the oil and eventually turns to acid if the airplane is not flown frequently. Ground running simply does not get the oil temperature hot enough. Prolonged ground running will result in hot spots, deteriorated ignition harness and brittle oil seals due to inadequate cooling. Only flying the airplane will ensure a high enough temperature is reached.

If the airplane is not going to be flown frequently, consult a licensed airplane maintenance engineer to determine the best action to take (such as changing the oil more frequently or changing the oil before parking the airplane for a lengthy period).

Monitoring Engine Performance

Every pilot should have an understanding of what the engine gauges and noise indications are telling him or her, so that any potential problems can be detected at the earliest possible stage. Attention should also be taken of any anomalies in performance that may indicate an engine problem (e.g. airplane speed, fuel consumption, oil consumption and engine power). Not all indications are clear-cut and different problems can have similar indications. Listed below are some typical indications a pilot may detect and some of the possible causes. Any indication that is not normal should not be ignored, as there is a reason for it.

Indication	Some Possible Causes
▸ Engine kicks back during start.	▸ Problem with the magneto impulse coupling. ▸ Out of specification engine timing.
▸ Engine will not start or is hard to start.	▸ Incorrect starting procedure. ▸ Too much fuel due to over priming (smell of fuel in exhaust pipe). Care needed because of fire potential. ▸ Fouled or defective spark plug. ▸ Water in the fuel system. ▸ Insufficient prime. ▸ Throttle open too far. ▸ Malfunction of start retard mechanism (detected by an absence of clicking with an impulse coupling or lack of buzzing with a "shower of sparks" system). ▸ Engine very hot (vapor lock).

Indication	Some Possible Causes
▶ Engine runs rough initially but then smoothes out after 3 – 5 minutes.	▶ Sticking valves.
▶ Engine runs rough during idle.	▶ Fouled spark plug(s). Either carbon fouling (rough running disappears when power increased) or lead fouling (rough running does not clear when power increased). ▶ Sticky valves (after engine warms, rough running may disappear). ▶ Incorrect idle mixture. ▶ Primer not locked, or is leaking. ▶ Air leak in the induction system. ▶ Faulty ignition system.
▶ Engine runs rough during run up.	▶ Lead fouled spark plugs. ▶ Engine not leaned at high density altitudes.
▶ Engine roughness during climb.	▶ Marginal magneto (if you can turn roughness off by reducing power).
▶ Engine roughness during cruise.	▶ Marginal or defective magneto. ▶ Carburetor icing. ▶ Induction system leak. ▶ Mixture too lean. ▶ Fouled or defective spark plugs.
▶ Engine fails to develop rated power.	▶ Restricted air inlet. ▶ Air leak in the induction system. ▶ Restricted fuel flow.
▶ High CHT.	▶ Cooling baffles missing, clogged, broken or improperly installed. ▶ Partially clogged injector. ▶ Engine improperly timed. ▶ Insufficient air cooling. ▶ Bird nest in engine compartment.

Indication	Some Possible Causes
▸ Low oil pressure.	▸ Insufficient oil. ▸ Excessive internal leakage (e.g. ruptured oil line/pressure relief valve operating out of adjustment). ▸ Incorrect grade of oil for the temperature (oil viscosity too low). ▸ High oil temperature. (Note that a sudden oil line rupture will not result in a sudden corresponding oil temperature rise.)
▸ High oil temperature.	▸ Insufficient oil. ▸ Incorrect grade of oil. ▸ Poor baffle seals (cooling air baffle missing, broken or damaged). ▸ Cooling air blockage. ▸ Leaking piston rings. ▸ Oil cooling system malfunction.
▸ Spark plugs continually become fouled.	▸ Excessively rich mixture setting. ▸ Worn rings. ▸ Improper spark plug heat range. ▸ Spark plug barrels that are dirty or wet. ▸ Ignition timing out. ▸ Improper leaning procedure. ▸ Rapid cool downs (e.g. low power descents).

Some other indications that require attention include:

> ▸ High manifold pressure at idle.
> ▸ Engine surges.
> ▸ Engine speed does not momentarily increase on application of idle cut-off (indication of incorrect mixture setting).
> ▸ Excessive manifold pressure for a given rpm.
> ▸ Low manifold pressure at full throttle (relative to altitude).
> ▸ Takeoff rpm lower than normal.

A common possible cause to all abnormal indications is a malfunctioning gauge.

Ground Operating Hazards

12

A surprising number of accidents occur while operating on the ground as the result of a collision with an object, or another airplane, while taxiing. A significant potential risk can also result from runway incursions in which an airplane creates a collision hazard, or loss of separation, with an airplane taking off, about to takeoff, landing or intending to land. *Both runway incursions and taxiing risks, are readily preventable.*

Taxiing

Although taxiing an airplane is a relatively straightforward operation, collisions with other airplanes (either moving or parked) happen all too often. Causes relate to inattention, taxiing too fast, breakdown in communication, proximity to larger airplanes, misuse of the brakes and poor judgement. Taxiing safely requires attention to the following:

- Do not move off without first looking for traffic and making sure the chocks and tie downs have been removed.
- Test the brakes just as you move off.
- Do not turn with one wheel locked. For short-radius turns, it is best to apply short jabs to the brakes, rather than locking up the brakes.
- Taxi slowly.
- Do not taxi with the brakes on.
- Keep a sharp look out, particularly when turning.
- Hold the stick back when taxiing on rough surfaces to keep the weight off the nose wheel (tricycle gear airplanes).
- Taxi very slowly when winds exceed 30% of stall speed.
- If crosswinds or gusts exceed 50% of stall speed, do not attempt to taxi without outside assistance.
- Ensure the ailerons and elevators are in the correct position relative to the wind.
- Be aware of any blind spots.
- If clearances do not look adequate, stop and get external guidance.

If something abnormal occurs, such as one brake seizes with a resulting sharp turn, or the brakes simply fail, then quickly pull the mixture to stop the engine. Make sure you do not relax your vigilance until the engine has been shut down.

Runway Incursions

While most runway incursions do not result in an accident, they do have the potential for serious consequences. A large number of runway incursions are the result of breakdown in communication, lack of airport knowledge, incorrect clearance interpretation, lack of understanding of the signs, confusion, disorientation or inattention. Fatigue, or lack of sleep, can increase the risk.

Reducing the risk of runway incursions can be achieved by:

- Being familiar with the relevant regulatory and local operating procedures applicable to taxiing, clearances and communication.
- Reviewing current airport diagrams during preflight planning (runways, taxiways and signs) and before taxiing and landing. Have the diagrams readily available in case of need.
- Reviewing NOTAM's for advice concerning any taxiway/runway closures, construction areas, or any other active constraints.
- Monitoring the correct frequencies and avoiding communication breakdown.
- Avoiding distractions and maintaining a sterile cockpit.
- Staying ahead of the airplane and maintaining a good lookout while the aircraft is moving.
- Being alert for other traffic (both airplanes and vehicles) and pedestrians.
- Stopping and asking for directions if lost or disoriented.
- Never making assumptions, and when in doubt, asking for assistance.
- Asking for progressive taxi instructions if unfamiliar with the airfield.
- Stating your position when calling for taxi instructions.
- Operating in accordance with ATC instructions, but seeking clarification when required.
- Maintaining a listening watch on the relevant frequencies, to visualize other traffic activity (e.g. monitoring ground and tower frequencies).
- Checking to ensure you are entering the correct runway.
- Checking for traffic before entering any runway or taxiway.
- At non-controlled airports, checking for traffic on any crossing runways.
- Staying alert, especially when visibility is poor or, the weather conditions are such, that surface markings are difficult to see.
- Making the airplane visible by using the landing lights in poor visibility.

Pilots, unfamiliar with an airfield, need to be particularly vigilant, especially at larger more complex airports. As stated by Transport Canada in *Take Five for Safety* – Look out, Listen out and Speak out.

Managing Health Risks

13

All pilots have a responsibility to ensure they satisfy the medical standards applicable to their licence. In addition, pilots must be knowledgeable about the many temporary health conditions that can impair their ability to fly, so that unnecessary risks are not taken. *Applying the acronym "I'M SAFE" is an effective way of carrying out a check on your level of fitness.*

Illness, Injury and Surgical Procedures

Flying with an illness, injury or after a surgical procedure, can have serious flight safety implications.

While some illnesses, injuries and surgical procedures clearly restrict your ability to fly, others can have less obvious affects. In addition, recovery can often take longer than you think. This is not necessarily helped by the fact that modern medical practices have reduced the periods of hospitalization and time off work. However, while this may allow you to return to your normal job at an early stage, it does not necessarily mean you are fit to fly. Where there is any doubt, consult your Aviation Medical Examiner.

Even short lived and minor illnesses, such as a cold or headache, will impair flying performance. For example, flying with a common cold can result in:

- An increased difficulty in clearing the ears during descent, leading to ear pain or possible eardrum rupture. Unblocking of only one ear can result in dizziness that can be severe and disabling.
- The semicircular canals in the inner ear being affected, leading to the increased risk of dizziness and spatial disorientation.
- Fatigue, and a general feeling of being unwell.

A headache can result in distraction, while a migraine can be incapacitating. And while some physical injuries may feel fine, they could be exacerbated when flying, due to inappropriate movement in the restricted cockpit environment.

It is therefore sensible not to fly until you are fully fit. If in doubt, check.

Medication

Pilot performance can be seriously downgraded by both prescribed and over-the-counter medications. Many prescription, and over-the-counter medications, can result in:

- Impaired judgment.
- Dulled alertness.
- Drowsiness.
- Moodiness.
- Poor coordination.
- Reduced reflexes.
- Poor balance.
- Disturbed vision.

There is also the potential for unexpected side effects and allergic reactions.

While the effects from medication may be inconsequential on the ground, they can compromise safety in a flying environment. It should be noted that over-the-counter medications usually only mask the symptoms, and do not necessarily provide a cure.

The following are some of the more widely used medications, which have side effects and interactions that could affect your ability to fly safely.

- Antibiotics.
- Tranquilizers, antidepressants and sedatives.
- Stimulants.
- Antihistamines.
- Certain drugs used to treat high blood pressure.
- Appetite and weight control medications.

It makes sense to heed the advice of the FAA Office of Aviation Medicine, who warns pilots to be wary of any illness that requires medication to make you feel better. They go on to note, that if the illness is serious enough to require medication, then it is also serious enough to prevent you from flying.

When uncertain, seek the advice of an Aviation Medical Examiner.

Dehydration

Dehydration occurs when the water losses from the body are not balanced by adequate water replenishment. High temperatures (such as a hot cockpit), a dry environment (such as a cold dry day), excess intake of drinks containing diuretic substances (such as caffeine) and other factors, can lead to this debilitating condition. Dehydration can seriously impair a pilot's performance with a resulting increase in flying risk. It can lead to impaired judgment, emotional alteration, impaired vision and reduced performance.

Symptoms of dehydration include:

- Thirst.
- Being unduly fatigued.
- Weakness and drowsiness.
- Feeling light-headed or dizzy.
- Impaired vision.
- Nausea.
- Headache.
- Slurred speech.

Note that thirst is an imprecise sign or indication of dehydration. It usually arrives too late and is turned off easily by a small amount of water – insufficient for rehydration. Avoidance is therefore the best policy.

Avoiding Dehydration

The primary prevention strategy is to maintain high body fluid levels by regular intake of suitable liquids such as water (preferably cool water). While each person is different, eight glasses a day is a guide. In extreme conditions, more will be needed.

- Don't rely on the sensation of thirst as an alarm signal. Drink before you become thirsty.
- On warm hot days, make sure you drink enough water before the flight.
- Have adequate water available in the cockpit during hot weather flying.
- Avoid beverages with caffeine or alcohol.
- Avoid fruit juices or non-diet soft drinks (sugar can aggravate dehydration).
- Avoid letting the cockpit get too hot.
- Open air vents before taxiing.
- Be aware that dehydration can be accelerated by vigorous exercise prior to flying.

Be particularly conscious of your general health and any pre-existing medical condition that may reduce your heat tolerance.

Carbon Monoxide

Carbon monoxide is the product of incomplete combustion and is present in exhaust gases of all internal combustion engines. It is colorless, tasteless and odorless. It is a potential danger in general aviation because many airplanes use the engine exhaust gas, through a heat exchanger, to warm the cabin. These heat exchangers can fail, resulting in leakage and exhaust gas entering the cabin.

The effect of carbon monoxide on the human body can be insidious, with confusion and unconsciousness occurring before the victim realizes that carbon monoxide poisoning is responsible. Breathing even minute quantities over a long period of time can have serious consequences. The best cure is prevention.

Causes of Carbon Monoxide Entering the Cockpit

Most airplanes have shrouds around the exhaust system to provide cockpit heating. If there is a crack or pinhole in the exhaust system, carbon monoxide can potentially enter the cockpit. Defects can consist of:

> - Worn or defective exhaust stack slip joint.
> - Exhaust system cracks or holes.
> - Openings in the engine fire wall.
> - Defective gaskets in the exhaust manifold.
> - Defective mufflers.

It is therefore prudent to prevent such defects from occurring by ensuring regular inspection of heaters and exhaust gas paths. As a backup, you can install "dead stop" patches (they change color in a carbon monoxide environment and are not expensive, but do need to be changed regularly).

Symptoms of Carbon Monoxide Poisoning

The presence of carbon monoxide can usually only be detected in airplanes by the associated engine exhaust smell (although dangerous amounts of carbon monoxide can be present even if no exhaust smell is detected) and/or the onset of symptoms such as:

> - Tightness across the forehead.
> - Headache.
> - Throbbing or pressure in the temples.
> - Vision problems.
> - Dizziness.
> - Confusion.
> - Nausea.
> - Drowsiness.
> - Ringing in the ears.
> - General weakness.

However, the symptoms can be subtle in that they are similar to other physiological problems. At its worst, carbon monoxide poisoning results in loss of muscular power, convulsions, coma and eventual death.

Action if Carbon Monoxide Poisoning is Suspected

▸ Immediately shut off all cabin heating.
▸ Change the cabin air source (open fresh air vent and open a window if possible).
▸ Avoid smoking.
▸ Declare an emergency and land at the nearest airfield.
▸ Obtain medical attention.
▸ Have the airplane checked for exhaust leaks before further flight.

Any time you smell engine type odors in the cockpit, or start to feel any of the symptoms, assume that carbon monoxide is present and act accordingly.

Stress

Stress is known to diminish a pilot's flying performance, and the combination of stress and fatigue can be extremely hazardous. For many people, however, stress is a topic that is ignored, denied or not adequately recognized. The issue is compounded by the fact that the mechanisms for coping with stress vary from person to person, and the effect on individuals is unpredictable.

The FAA Advisory Circular AC 60-22 *Aeronautical Decision Making*, defines stress in the following terms:

"Stress is the term used to describe the body's non-specific response to the demands placed on it, whether these demands are pleasant or unpleasant in nature. Stress is an inevitable and necessary part of life that adds motivation to life and heightens a pilot's response to meet any challenge. In fact, performance of a task will generally improve with the onset of stress, but will peak, and then begin to degrade rapidly, as stress levels exceed a pilot's adaptive capabilities to handle the situation."

Therefore, while some stress is beneficial, too much stress can lead to a situation that is beyond the pilot's ability to respond appropriately.

Impact of Stress on Flight Safety

The impact of stress on flight safety can be significant, depending on the individual and the particular circumstances. For example, excessive stress can lead to:

> - Reduced performance that can be aggravated by a high cockpit workload. The task requirements may eventually exceed the pilot's capability.
> - A focus on one problem to the exclusion of all other tasks.
> - A decrease in the ability to absorb all the information needed for the safety of the flight.
> - Distraction and loss of mental awareness, as the result of preoccupation with non-flying matters.
> - A tendency to react automatically on the basis of previous well learned routines, which could be inappropriate to the airplane being flown.
> - Deterioration in the ability to make error-free judgments and decisions.
> - A propensity to take risks which, under normal circumstances, would not be taken.

The result is a reduction in the margins of safety, with a consequential increase in the risk of an accident. That is, an increase in the potential for errors, and less likelihood of being able to break the error chain.

Causes of Stress

Stress can result from matters unrelated to flying (life induced stress) such as:

> - Financial worries.
> - Loss of sleep.
> - Missed meals.
> - A sick child.
> - Matrimonial argument.
> - Interpersonal conflict.
> - Work related stress.

Flying itself can result in stress (aviation induced stress) depending on circumstances. For example:

> - Not adequately knowing some aspects of operating the airplane.
> - Inadequate proficiency.
> - Abnormal conditions.
> - Being lost.
> - Excessive workload.
> - Being rushed.

All pilots need to be aware of the potential for both types of stress, recognize the symptoms, and take action to minimize adverse stress.

Symptoms of Not Coping with Stress

There are numerous signals that indicate a person is not coping with stress. They can be emotional, physical or behavioral in nature and include such things as:

- Depression.
- Preoccupation.
- Withdrawal.
- Insomnia.
- Anger and hostility.
- Loss of control.
- Irritability.
- Impatience.

General Life Stress Reducing Mechanisms

Some of the common ways to reduce stress, caused by conditions unrelated to flying, are:

- Physical activity.
- Leisure activity.
- Positive self-reinforcement.
- Good relationships with people.
- Relaxation activity.
- Taking breaks.
- Eating balanced meals.
- Avoiding caffeine and sugar.
- Getting adequate sleep.

Aviation Stress Reducing Mechanisms

There are also numerous ways to minimize stress that can potentially be generated while flying.

- Ensuring your proficiency level meets the needs, and potential needs, of the flight.
- Carrying out thorough and unhurried preflight planning and inspection.
- Minimizing the cockpit workload by:
    - Maintaining situational awareness.
    - Staying ahead of the airplane.
    - Thorough preflight planning.
- Avoiding distractions.
- Being prepared for abnormal situations.
- Knowing and respecting your own limits.
- Not letting mistakes bother you, while similtaneously being ready to recognize them.
- Asking ATC for help when necessary.
- Keeping to your personal minimums.
- Knowing your airplane well.

There are clearly many things that can be done to reduce stress, as well as improve your ability to cope with stress. In particular, the better prepared you are for flying, the better equipped you are to handle abnormal situations that may arise. However, in the end, if you are suffering from a high level of personal stress and/or are simply not properly prepared – don't fly.

Fatigue

Fatigue continues to be a major hazard to flight safety. It is especially dangerous because it affects people before they are even aware of it. In addition, as pointed out by Transport Canada in Issue 4/82 of the *Aviation Safety Letter*, fatigue "doesn't care who you are, your experience, or the type of aircraft flown. It feeds on dedication, ambition, greed, over-confidence, not knowing your own limits, and a reluctance to say "enough" ".

Causes of Fatigue

There are many causes of fatigue, some of which are:

- Lack of sleep.
- Continuous high demands on physical and/or mental energy.
- Hunger or thirst.
- A poor cockpit environment (extreme temperature, noise, vibration, poor ergonomics).
- Pressures (either self-induced or from others).
- Stress.
- Alcohol, drugs or heavy smoking.
- A poor state of health.
- Dehydration.

Consequences of Fatigue

The consequences of fatigue are highly dangerous when flying. For example, fatigue can result in:

- Errors in judgment.
- Impaired decision making.
- Slower reaction times.
- Reduced motivation.
- Reduced ability to concentrate.
- Inattention.
- Increased likelihood of errors.
- Neglect of deviations.
- Increased irritability and lack of patience.
- Decreased self-discipline.
- Reduced vigilance (e.g. ignored hazardous warning signs).
- Acceptance of lower standards of accuracy and performance.
- Perseverance with ineffectual solutions.

Some Signs of Fatigue

> ▸ Difficulty in staying awake and frequently yawning.
> ▸ Poor short-term memory recall.
> ▸ Forgetting to carry out some routine actions.
> ▸ An inability to maintain flying tolerances (e.g. altitude and heading).
> ▸ Wandering and disconnected thoughts.

How to Avoid Fatigue

The first step is to be able to recognize when you are at risk and avoid attitudes that result the in signs of fatigue being ignored. In addition, take the following preventative measures:

> ▸ Get adequate sleep before a flight.
> ▸ Have regular sleeping and eating habits.
> ▸ Avoid alcohol and don't depend on it as a sleep aid.
> ▸ Have a proper diet.
> ▸ Drink adequate amounts of water.
> ▸ Avoid junk food.
> ▸ Keep the body in good physical condition.

It should be remembered that when you are fatigued you would probably accept, without being aware of it, greater risks. This is potentially dangerous, and it is therefore sensible to avoid becoming fatigued in the first place. If the signs of fatigue are there - don't fly, or, if you are already flying, land at the nearest airport.

Hypoxia

A shortage of oxygen in the body leads to a condition known as hypoxia. When flying, hypoxia results from the reduced partial pressure of oxygen as altitude increases, with a resulting deficiency of oxygen in the blood stream. Hypoxia leads to impairment of various organs in the body including the brain. In general, hypoxia begins to take effect above 10,000 feet but can be at lower altitudes depending on the rate of ascent, time spent at altitude, fatigue, stress, level of fitness and how efficient the body takes in oxygen. In addition, if a person is a smoker, is taking medication or is unhealthy, then he or she is more susceptible to hypoxia.

Symptoms of Hypoxia

The onset of hypoxia generally occurs slowly and insidiously. The signs and symptoms may develop so gradually that they may be well established before you even recognize them. You may, in fact, experience a feeling of "well being" resembling mild intoxication, leading to a false sense of security that also makes detection difficult. Detection is compounded by the fact that symptoms vary in intensity and sequence from individual to individual. The following symptoms can therefore occur without the pilot recognizing the condition:

> ▸ Subtle impairment of vision.
> ▸ Increased breathing rate.
> ▸ Ineffective self-checking.
> ▸ Impaired mental capability.
> ▸ Clumsiness.
> ▸ Drowsiness.
> ▸ Light headedness or dizziness.
> ▸ Increased propensity for errors.

The longer you remain above 10,000 feet the worse the effects will become, with the rapidity of symptom onset depending on the altitude. The higher the altitude, the greater the rate of onset, with time to useful consciousness being between 20 and 30 minutes at 18,000 feet. Even at 15,000 feet, a pilot's performance can seriously deteriorate within 15 minutes. Worsening symptoms include:

> ▸ Tingling feeling of the skin.
> ▸ Increased heart rate.
> ▸ A dull headache.
> ▸ Blue lips and nails.
> ▸ Fuzzy vision.
> ▸ Tunnel vision.
> ▸ Eventual unconsciousness.

Because hypoxia is so difficult to detect, the best solution is to prevent its occurrence.

Preventing Hypoxia

There are two simple ways to avoid getting caught by the effects of hypoxia.

> ▸ Do not fly above 10,000 feet unless you have, and use, supplemental oxygen. Note that night vision can deteriorate at altitudes as low as 6,000 feet, and it is therefore recommended that supplemental oxygen be used above this altitude when flying at night.
> ▸ Do not gauge oxygen starvation by how you feel.

Remember that some people can be affected by hypoxia below 10,000 feet and suitable caution needs to be exercised. Spending time in an altitude chamber can help identify your susceptibility. Don't let a hazardous attitude cause you to violate the above preventative strategies.

Handling Emergency Situations

14

Emergencies such as an engine failure, a complete electrical failure or even a precautionary landing, are not everyday occurrences. Hence, our readiness for these situations is questionable unless we avoid an "it will never happen to me" mindset, and regularly learn and practice simulated events. *A pilot's ability to survive an emergency depends upon training, knowledge, skill, and the ability to properly analyze the situation.*

What follows, is general guidance on addressing emergency situations. In all cases the POH should be referenced for the emergency procedures applicable to the particular airplane.

Precautionary Approach and Landing

Under certain adverse circumstances a precautionary landing may be necessary when the hazards of continued flight are greater than those associated with a precautionary landing. A controlled precautionary landing is always a safer option when compared to a subsequent engine-out forced landing.

Possible Reasons for a Precautionary Landing

- You are lost and low on fuel.
- Low oil pressure or other abnormal engine gauge reading is detected. (It could be a faulty gauge, so check other instruments.)
- You are short of fuel and there is no nearby airfield. (It is useful to plot country airstrips on your map along your planned route in case of emergency.)
- A significant oil leak. (Don't leave the oil filler cap off during preflight inspection.)
- There is little daylight left and you are not night proficient.
- There is an electrical fire or smoke from an electrical fault.
- When bad weather, such as a low ceiling, is encountered with no escape route.
- A serious airframe or power plant vibration suddenly occurs.
- A partial engine power loss occurs.

Landing Sites

A precautionary landing offers a better choice of landing sites compared to the situation when your engine fails suddenly. While it is not without risk, sensible selection of the landing site increases the chances of survival. Note that roads are generally not good landing sites because of power lines, sign and vehicles.

Precautionary Landing Procedure

The procedure for carrying out a precautionary approach and landing will vary with the circumstances. The procedure below is one option.

- Slow down the airplane and deploy the flaps. (Increases time to assess the situation.)
- Look for a suitable landing area.
- Transmit your intentions and give your position.
- Circle the intended landing area at a safe altitude and identify any obstructions, wires, poles, dead trees etc.
- Determine the wind direction and velocity (e.g. note smoke or drift).
- Confirm the acceptability of the landing area by carrying out an inspection run at 200 feet AGL (or well above possible obstructions), into wind, and slightly to the right of the intended landing area. Check out the following:
    - The approach.
    - Landing area length (by timing from one end to the other).

> - Landing area slope.
> - Texture and condition of the surface.
> - Possible wind shear due to trees.
> - Wind conditions.
> - Landing area obstacles such as animals, holes, rocks, temporary fences, equipment and crossing tracks.
> - Obstacles, both on approach, and possible over shoot.
> - The touch down area.
> - Go around and carry out a practice short field approach in the approach configuration to get a feel for the conditions. Overshoot at not lower than 50 feet AGL and carry out a further inspection of the landing area.
> - Carry out another circuit and land.

Whether to carry out a precautionary landing or not, is a matter of judgment depending on the options available. It is important, however, to always be mentally prepared for it as a possibility, so that a timely decision can be made, if the risk of continued flight is less desirable.

Forced Landings

Engine failure due to mechanical problems are relatively rare these days, provided the engine is maintained and correctly operated. However, engines can fail due to pilot related causes, the most common of which are fuel starvation, fuel exhaustion or induction system icing - all of which are preventable. Good preflight planning and inspection, coupled with the use of checklists, will go a long way to minimizing the risk of engine failure due to these causes, and are discussed elsewhere in this manual.

However, should a failure occur for whatever reason, it is vital that the pilot be mentally prepared to act quickly, be proficient at the forced landing procedure and have the required flying skills to handle the prevailing conditions. Therefore periodic forced landing practice with an instructor, under a variety of conditions, is advisable.

In dealing with engine failure, there are a number of tactics and considerations that will improve your chances of surviving a forced landing. These are discussed below and are supplementary to the emergency landing procedures in the POH and relevant training manuals.

Preflight Planning

A pilot's ability to execute a safe off-field landing can be improved by following some simple guidelines during preflight planning.

> ▸ Avoid rough, heavily treed, hilly or mountainous terrain. If there is no alternative, choose a route that takes you over possible emergency landing areas.
>
> ▸ Plan to fly at the highest practical altitude above the ground. (The higher you are the more options you have.)
>
> ▸ Avoid flight over extensive areas of fog.
>
> ▸ Be ready for an engine failure before every flight (procedure recall and proficiency).
>
> ▸ Be aware of the topography along the intended route and consider what emergency landing areas are available.
>
> ▸ Identify possible landing sites when planning a flight over populated areas.
>
> ▸ Be aware of the winds aloft and winds at ground level.
>
> ▸ Be familiar with the best glide speed and stall speeds of the airplane.

In addition, if flying at night, or in IMC:

> ▸ Avoid areas of low ceiling and visibility.
>
> ▸ Plan a route that takes you over as many en-route airfields as possible. (Engines generally give some warning of impending failure, and a precautionary landing can therefore be more readily achieved.)
>
> ▸ Avoid extensive areas of unlit terrain.
>
> ▸ Be prepared to follow your route on a suitable chart so you will know your position and the nature of the emergency landing options.
>
> ▸ If possible, fly on a clear night with some moonlight.

Being Ready for an Emergency

To be able to handle an emergency situation effectively, it is important that you:

> ▸ Always be ready for an emergency (i.e. consider it as a possibility and constantly be aware of where you might put down in case of an emergency). Carry out a power off landing every now and then to maintain gliding proficiency.
>
> ▸ Accept the emergency situation and act without delay (i.e. don't let your mind become "paralyzed").
>
> ▸ Avoid any desire to save the airplane.
>
> ▸ Maintain composure and avoid panic.

Engine Failure on Takeoff

The generally accepted rule is, that if the engine fails during takeoff, land straight ahead and don't turn back to the runway. Turning back below 1000 feet is a high-risk proposition, while landing straight ahead in a controlled descent, is much more survivable than a stall/spin - hence the rule.

However, as you gain height you have an "expanding cone of options". The minimum altitude at which a return to the airfield can even be contemplated will vary from pilot to pilot. It requires rapid reaction time, an ability to execute a 3 x standard rate turn without stalling, knowledge (and experience) of the height loss characteristics of the airplane, the skill to handle a downwind landing and a runway with adequate length. For these reasons, a specific altitude cannot be stipulated, and for most pilots, the safest action is to simply land straight ahead, making gentle turns as required within the cone of options.

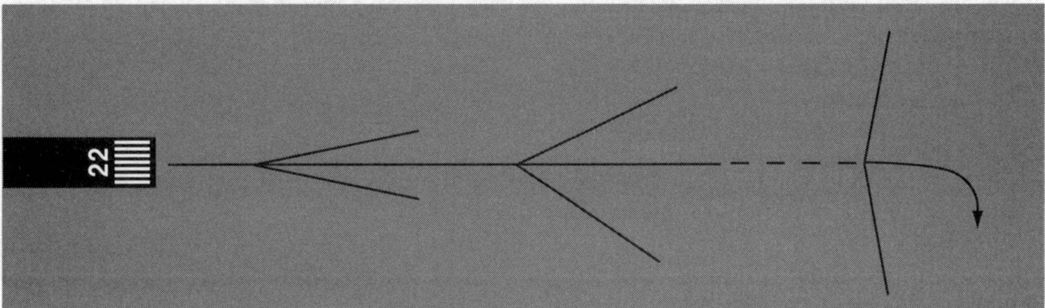

Engine Failure in Cruise

The immediate actions following an engine failure are to establish the best glide speed, select a suitable landing area, and plan the approach. Selection of the landing area requires judgments to be made about the gliding distance that can be achieved, the acceptability of the surface and the adequacy of the length. While gliding distances are given in the POH, a useful rule of thumb for most light airplanes is to use one nautical mile for every 1,000 feet above ground level. Note, however, that the direction and speed of the wind, impacts on the gliding distance that can be achieved. Knowledge on how to maximize the gliding distance with either a tailwind or a headwind will extend the possible options. In any case, gliding proficiency and judgment is helped by regularly carrying out power-off approaches.

Ron Fowler in his book, *Making Perfect Landings*, suggests that selecting a landing area that lies within a 45-degree cone, can minimize the need for judgment and experience. "By confining your landing area to the cone you are assured of reaching it, even against a moderate wind, towards up sloping terrain or with a poorly flown airplane" Fowler writes.

If distance to a landing area is not an issue, then flying at 10 to 15 knots above the stall speed can maximize time. For airplanes that have a constant speed propeller, moving the propeller control to low rpm/course pitch will reduce drag.

When choosing a landing site, color can be used to provide information about the nature of the surface.

> ▸ Brown with a touch of green is best (e.g. stubble field).
> ▸ Dark green is a poor choice (could be wet and soft).
> ▸ A newly ploughed field is normally a poor choice but if there is no option, land in the direction of the furrows.

Part of the field selection process necessitates looking for obstacles such as power lines, telephone lines, ditches, fences and stock. While an into-wind landing is preferable, obstacles may dictate a downwind or cross wind landing.

Landing

If you are faced with an emergency landing, it is useful to remember that injury can be avoided, or minimized by:

> ▸ Keeping the cockpit/cabin relatively intact.
> ▸ Using dispensable parts of the airplane, such as the wings, to absorb some of the forward momentum.
> ▸ Avoiding forceful contact with the interior of the cockpit (i.e. be restrained).
> ▸ Landing at the lowest possible ground speed.
> ▸ Landing under control.

If faced with a treed landing location, land between the trees, as this action will help slow the airplane (even though the wings may be sheared off) before contact is made with the ground. If you stall above the trees, then in all likelihood, you will hit the ground vertically.

In confined areas, it may be preferable to force the airplane onto the ground rather than delay touchdown until the airplane stalls. Ground looping is another option. If the decision is made to land on a road, bear in mind the inherent risks associated with possible traffic, power lines or overpasses. Ditching requires knowledge of the ditching procedures.

Whether the landing gear should be up or down, depends on the terrain and the gear position stipulated in the POH. However, note that if the landing gear gets torn off, it could severely damage the fuel tanks in a low wing airplane.

Emergency Procedure Recall

One useful memory jogger is to use acronyms. Another technique is to use a flow pattern where the most likely cause of engine failure is tackled first, and the remainder is acted upon as you move in a definite pattern through the cockpit. Whatever system is used, practice is required.

Some General Points

Here are some general points to remember:

> ▸ Concentrate on flying the airplane and landing safely as a priority. Attempts to troubleshoot the cause of the emergency should be made only if time permits.
> ▸ If the engine failure occurs at considerable altitude, initially select a general landing area, and later the specific site.
> ▸ Plan to reach the intended landing area with excess altitude.
> ▸ Keep the selected landing site in view.
> ▸ Be cognizant of the wind direction and strength, as it can have a significant impact on the gliding distance.
> ▸ Do not bleed off excess height until a landing is assured.
> ▸ Do not stretch the glide.
> ▸ Side slipping can be used to reduce height without an increase in forward speed. However, proficiency at this maneuver is required, and the airplane limitations of such a maneuver understood.

Ditching

Flying over a large expanse of water in single engine aircraft is generally permitted subject to certain equipment (e.g. life jackets, life raft) being carried depending on the nature of the flight. This type of flying requires:

> ▸ Extra preflight planning.
> ▸ Careful instructions to passengers.
> ▸ Appropriate emergency equipment being carried.
> ▸ Accessibility of emergency equipment.
> ▸ Knowledge of how to ditch and subsequent survival.

Preflight Planning

> ▸ It is not advisable to fly over a large body of water at night, as a successful ditching will be difficult to achieve.
> ▸ It is a good idea to route the flight over islands if possible.
> ▸ Don't be lulled into complacency – "it is only 50 nautical miles".
> ▸ Make certain a flight plan has been filed and that the flight proceeds on a reporting schedule.
> ▸ Avoid flying over water where the temperature of the water is less than 15°C as you cannot survive for long in anything less.
> ▸ Review your knowledge of how to use the emergency equipment.
> ▸ Review your knowledge of the aircraft emergency egress procedure.
> ▸ Review your knowledge of the ditching procedures and survival techniques.

Emergency Equipment

In addition to life jackets, consideration needs to be given to carrying the following:

- High quality inflatable life raft.
- Signalling mirror.
- Flare pistol.
- Dye marker.
- Hand held transceiver in a sealable plastic bag
- Flotation type portable ELT.
- Pocketknife.
- Bailing bucket.
- Life raft patch kit.   }
- Limited supply of rations.   }   If you are carrying a raft.
- First aid kit.   }

All emergency equipment should be readily accessible. Signalling devices, portable ELT, portable transceiver should be carried on the pilot's body or attached by lanyards.

Passenger Brief

Passenger briefing is mandatory and it is important that this be done thoroughly. The briefing should cover:

- Donning and use of life jackets
  (it is probably a good idea to wear them while over water).
- Ditching procedure and what to expect.
- Egress procedure and everyone's roles.
- Use of emergency equipment.

Ditching and Survival Procedures

Ditching requires a thorough knowledge of the recognized ditching and survival procedures and is too extensive to cover here. Suffice to say that there are numerous excellent publications on this subject, some of which are listed in the Selected Bibliography at the end of the manual.

Electrical Failures

Electrical problems during flight do occur, and while most failures do not lead to accidents, analysis shows that pilot error, in responding to the situation, is not uncommon.

The implications of a complete electrical failure depend on the circumstances and the pilot's preparedness to handle the ensuing situation. Flying VFR poses the least difficulty, although loss of power can leave you without flaps and the need to manually raise or lower the landing gear. Communication can also be lost once the battery power has been expended. Total electrical failure in a complex twin engine airplane, during an approach in IMC, with only one pilot on board, obviously results in a much more serious situation.

Electrical problems need to be handled correctly and promptly because of the possibility of electrical fire, or damage to electrical equipment. Alternatively, the problem may only be spurious in nature and recognizing this, can avoid a difficult situation. A thorough knowledge of the airplane electrical system and the electrical failure procedure is therefore essential. In particular you need to:

▸ Have knowledge of which equipment and instruments are electrically powered.

▸ Be able to recognize the indications of electrical failure or malfunction. Know which type of ammeter is installed (zero-center or zero-left) and know what the indications mean.

▸ Have knowledge of the procedures for responding to over voltage and under voltage indications.

In addition, your ability to handle any subsequent situation safely requires:

▸ Knowledge of the non-essential electrical loads.

▸ Knowledge of how to shed non-essential electrical equipment.

▸ The availability of an operating flashlight when flying at night, and proficiency at using it while flying. (Nowadays, flashlights that hang around your neck and leave both hands free, are available.)

▸ Proficiency at flapless landings.

▸ Knowledge of how to manually extend the landing gear on a retractable gear airplane.

▸ Proficiency at flying in IMC without electrical power.

Failure of an individual item is usually accompanied by a circuit breaker "popping". There is usually a reason why circuit breakers "pop". Hence it is advisable that they only be reset if it is essential to have the system operating. As with forced landings, periodic practice under simulated conditions, and knowledge of the relevant procedures will help prepare you for any possible electrical failure.

Instrument Failures

All pilots should have some partial panel training and be able to identify the indications of instrument failure. However, there is a significant difference between a training exercise and a situation where you not only have to identify a failure (which is not always obvious), but have to overcome the natural tendency to accept, as correct, the information supplied by a failed instrument. Most accidents are the result of not recognizing the problem in the first place, so the ability to detect instrument failure is clearly important.

Instrument Failure Detection and Diagnosis Training

In order to detect instrument failure, it is wise to obtain appropriate training. This can be achieved through the use of a simulator, or personal computer-based aviation training device, so that random slow failures can be simulated. Such training is invaluable.

Detecting Instrument Failure

Sudden, drastic failure of an instrument, or power supply system, is relatively easily discernible, but a slow malfunction is harder to detect. It is therefore necessary to:

> ▸ Know what drives each instrument (vacuum, electric, static air, pitot tube).
> ▸ Know the common failure modes.
> ▸ Cross check instruments during the normal scan, including when flying on autopilot. (Most autopilots are coupled to the attitude and heading indicators.) It is advisable to pre-plan the method of determining whether an instrument has failed or not, so that you are ready to diagnose any unusual indication.
> ▸ Include the vacuum gauge and ammeter in your scan.
> ▸ Check the heading indicator against the compass every 15 minutes, and if the error is large, it could mean a vacuum system failure.

Vacuum Pump Failure

Most instrument failures result from a loss of vacuum. Vacuum pumps, despite their importance, have been a source of continual premature failure, and can fail without warning. The system can also fail slowly, due to a dirty filter, a sticking regulator, a worn out pump or a leak in the system. Malfunction of a single-pump vacuum system will result in loss of the attitude indicator, the heading indicator, and probably the autopilot, depending on the instrument configuration. The loss of the vacuum system in IMC, without a backup system, can develop into a hazardous situation unless the problem can be diagnosed, and the airplane flown by partial panel into visual meteorological conditions. Alternatively, it may be necessary to carry out an approach without gyroscopic instruments, and practice at this procedure, under simulated conditions, is advisable.

Airspeed Indicator Failure

The airspeed indicator can fail (i.e. misread) if the pitot tube becomes blocked due to:

> ▸ Insects.
> ▸ Ice formation resulting from pitot heat failure. (It is important, therefore, to physically check that the pitot heat is functioning properly during preflight inspection.)
> ▸ Ice formation resulting from not switching on pitot heat.

If you don't have a warning light, pitot heat malfunction can be difficult to detect. The only way is to sense whether the airspeed reading is correct or not.

Pitot heat should be switched on in advance of flying in cloud or precipitation (regardless of the outside air temperature), and in high humidity when the temperature is at or below freezing.

A blocked static tube can also cause the airspeed indicator to misread.

Early Warning Signs of Possible Instrument Failure

The gyroscopic instruments can give the pilot early warning signs of impending failure.

> ▶ Listen to the gyro instruments during start up and shut down. If growling or grinding noises are heard, or a particular instrument stops quickly, then the instrument will need to be checked.
>
> ▶ If the attitude indicator erects slowly, or oscillations don't dampen out quickly (or it doesn't oscillate at all) then have the instrument checked.

Recommended Actions for Instrument Failure

If there is an instrument failure, it is clearly important that you know the secondary instruments and have the required level of proficiency at partial panel flying. It is also useful to know the engine power and propeller settings for climb, cruise, descent and approach, to assist the partial panel flying operation (e.g. when you have an air speed indicator failure).

The first action is to maintain control of the airplane (aviate, navigate, communicate). Then:

> ▶ Cover the failed instrument(s) if possible.
> ▶ If the failure does not affect the autopilot, use it. Otherwise, turn it off.
> ▶ Use alternative instruments in place of the failed instruments.
>> ▶ An ADF with a compass card can replace the heading indicator (if within range of a station).
>> ▶ A GPS can provide almost instantaneous updates of heading during a turn.
>> ▶ A GPS can also be used as an airspeed guide by noting the ground speed and adjusting for wind speed (in case the air speed indicator fails).
>> ▶ A GPS can provide altitude readout in case of an altimeter failure.
> ▶ If you have a heading indicator failure (and no GPS) use timed turns.
> ▶ Inform ATC of the problem.

If you fly IFR in single engine airplanes, it is recommended that a backup vacuum system be installed, or alternatively, provide dual attitude indicators, one vacuum driven and one electrically driven.

Fires

Fires are extremely rare, but in the event of an occurrence, quick action is needed. Procedures for addressing various fire scenarios are given in the airplane POH although there is usually a note to the effect that the procedures are "general in nature and pilot judgment should be the determining factor". However, pilot judgment will depend on:

> ▸ Knowledge of the airplane and its systems.
> ▸ How recently a fire emergency was practiced.
> ▸ The extent to which the type of fire can be recognized.
> ▸ The level of understanding of the reasons behind the procedural steps.
> ▸ The degree to which the procedure can be recalled.

The ability to react quickly and appropriately is vital in effectively dealing with a fire emergency. Gaining that ability requires periodic practice and knowledge refreshment.

Avoiding Fires

The best way to avoid an airplane fire is careful maintenance and thorough preflight inspection. When doing the walk around:

> ▸ Check for evidence of fuel, oil or hydraulic leaks.
> ▸ Check the exhaust system for security and leaks.
> ▸ Check the electrical equipment.

Engine Fire During Start

A fire during engine start generally results from over-priming in cold weather when excess fuel flows into the carburetor intake, with backfiring causing ignition. The general procedure is to keep cranking the engine and switch off the fuel. This should suck the flames back through the carburetor and extinguish the fire. Check the POH for the procedure applicable to the particular airplane.

In-flight Electrical Fires

An in-flight electrical fire can usually be detected by an odor of burning insulation, or the presence of white smoke (although white smoke can also result from a fuel fire burning certain material under the cowling). Causes can be over voltage, higher than normal electrical loads or a failed electrical component.

The exact procedure depends on the airplane, and reference should be made to the POH. However, in general, the actions are:

> ▸ Continue to fly the airplane.
> ▸ Turn off the master switch.
> ▸ Close the air vents.
> ▸ Use a halon fire extinguisher if appropriate.
> ▸ Take action to manage the smoke in the cabin. Various manufacturers recommend different ways of addressing this problem, with possible options including:
> > ▸ Changing the wing flap setting to alter the flow of air.
> > ▸ Opening a window or door away from the pilot, although this can well exacerbate the fire.
> > ▸ Side slipping the airplane to change the airflow.
> ▸ Turn off all electrical equipment.
> ▸ If possible leave the master off. If, however, an electrical system is needed for safety reasons (e.g. landing gear), then turn on the battery switch first and check for a fire indication. If there is none, then turn on the alternator and check again.
> ▸ If the circuit breaker of a necessary circuit has popped, leave it off until near the airfield.
> ▸ Land as soon as practicable.

In-flight Fuel or Oil Fires

Engine fuel or oil fires are generally the result of engine component failure due to age, fatigue, damage or improper maintenance, resulting in fuel or oil being ignited by the hot engine, or an errant spark. Recognizing whether an in-flight fire exists may not be that easy. Possible symptoms include smell, black smoke, a rough running engine or a drop in fuel pressure.

The exact procedure depends on the airplane, and reference should be made to the POH. However, in general, the actions are:

> ▸ Continue to fly the airplane.
> ▸ Set the mixture control to idle cut off.
> ▸ Turn off the fuel/fuel pump.
> ▸ Close the air vents and manage the air in the cockpit as per the electrical fire procedure noted above.
> ▸ If appropriate, carry out a rapid descent that may blow out the fire.
> ▸ Land as soon as possible.

Regardless of the type of fire, they are extremely dangerous and an immediate landing is required.

Propeller Failures

Separation of part, or all, of the propeller blade is not an emergency we are normally trained to deal with, nor generally do we even contemplate it happening. However, you should always be prepared in case a failure, or other severe vibration occurs. Propeller blade failure can be followed very quickly by failure of the engine mounts, and possible separation of the engine from the airplane. If failure occurs, a rapid response is clearly required.

Propeller Blade Failure

Blade failure is easily recognized by either severe vibration or engine roughness. Note that the latter could be due to other causes, such as induction icing, failure of one cylinder or crankshaft failure. The immediate action is to reduce the throttle to idle, and shut down if you are sure a propeller blade failure has occurred. Unless you are certain that the problem is something other than propeller failure, do not advance the throttle during the forced landing cause check.

Propeller Governor Failure

Governor failure can be due to a broken oil line (identified by dropping oil pressure and increasing temperature), or a spun bearing or internal leakage that may restrict oil flow to the propeller. It is recognized by engine over-speed and/or poor propeller control (engine over-speed is indicated by wild changes in propeller rpm). The action is to immediately reduce the throttle to idle and set up for a forced landing. During the cause check, determine if the oil pressure and oil temperatures are in the green. If so, slowly advance the throttle to assess the reaction. There may be a setting at which the propeller settles down and you can return to the nearest airfield.

Appendix 1
Typical Airplane
Knowledge Checklist

Operating Speeds

- Normal rotation speed.
- Best angle of climb speed.
- Best rate of climb speed.
- Normal climb speed.
- Normal cruise speed.
- Maneuvering speeds (at various weights).
- Never exceed speed.
- Maximum structural cruise speed (top of green arc).
- Maximum landing gear operating speed.
- Maximum landing gear extended speed.
- Maximum flap extension speed (may be different for different settings).
- Stall speed – no flaps.
- Stall speed – full flaps.
- Stall speed – full flaps/gear down.
- Stall speed at 60° angle of bank.
- Normal descent speed.
- Best glide speed (max. weight).
- Glide ratio.
- Best speed for range.
- Best speed for endurance.
- Landing speeds.
    - Base.
    - Final.
    - Short field (full flap).
    - Flapless.
- Normal cruise speed.
- Normal operating speed range.
- Maximum demonstrated crosswind.
- Calibrated airspeed versus indicated airspeed.

Emergency Procedures

- ▸ Engine failure.
- ▸ Electrical failure.
- ▸ Vacuum failure.
- ▸ Instrument failure.
- ▸ Landing gear failures.
- ▸ Smoke and fire.
- ▸ Order of instrument shedding in case of electrical failure.
- ▸ Dealing with a "popped" circuit breaker.
- ▸ Estimated battery life in case of electrical failure.
- ▸ Propeller system failure.

Electrical System

- ▸ Location of critical fuses for landing gear, flaps, landing lights, generator/alternator.
- ▸ Type of ammeter gauge, and what the indications denote.
- ▸ Electrical annunciation lights, and what they mean.
- ▸ How to detect an electrical failure.
- ▸ Basic understanding of the system.
- ▸ Battery amp hour rating.
- ▸ Instrument and equipment power requirements, and emergency load items.
- ▸ Full charge ammeter indication.

Fuel System

- ▸ Location of all drains.
- ▸ Usable fuel capacity in each tank (full/tabs).
- ▸ Fuel system understanding.
- ▸ Fuel management procedures (draining, filling, sampling).
- ▸ Nominal fuel consumption rate.
- ▸ Correct grade of fuel.
- ▸ Auxiliary feed pump operation.
- ▸ Fuel tank switching procedure.
- ▸ Fuel tank selector switching system.

Engine

- Basic engine configuration and systems.
- Engine starting procedures (normal, hot, flooded).
- Oil quantities (maximum, minimum).
- Required type and grade of oil.
- Type of fuel control (carburetor, fuel injection).
- Engine leaning procedure.
- Procedure for use of carburetor heat/alternate air.
- Type of magneto start retarding mechanism.
- Power setting(s) for different percent power.
- Propeller operation (constant speed).
- Normal gauges readings.
- Induction system and alternate air operation.

Instruments

- Which instruments are vacuum powered and which are electrically powered.
- Number of vacuum pumps and method of activation.
- Indication of loss of vacuum.
- Alternate static source and method of activation.
- Which instruments the auto pilot uses and the consequences of failure of those instruments.
- Communication equipment function/use of squelch etc.
- Testing procedures (e.g. navigation aids, autopilot).
- Location and actuation of alternate static air.
- Autopilot operating limits.
- Operation of all instruments including GPS, autopilot, communication, head sets/intercom, transponder etc.
- Heater function and controls.

Landing Gear

- Procedure for lowering/raising.
- Emergency procedure for lowering the gear.

Weight and Balance

- How to check weight and balance using the data provided.
- Loading limitations (fuel, luggage, and/or passengers).
- Airplane basic empty weight.
- Airplane maximum landing weight.
- Airplane maximum takeoff weight.

Takeoff and Landing

- ▸ How to check takeoff distance, landing distance, climb gradients and weight limitations, using POH and other data.
- ▸ Required tire pressure.

Normal Procedures

- ▸ All normal procedures.

General

- ▸ Location, use and operation of the fire extinguisher.
- ▸ Details of any modifications on the airplane.
- ▸ Speed in reduced visibility.
- ▸ All airplane limitations.

Appendix 2 Preflight Planning Checklist

A2

Personal Preparedness/Readiness

- Have I allowed sufficient time to avoid rushing?
- Have I adequately assessed all the risks?
- Am I current (medical, recency etc)?
- Am I fit to fly ("I'M SAFE")?
- Am I completely familiar with the particular airplane and its operating procedures?
    - Equipment operation.
    - Emergency procedures.
    - Fuel systems.
    - Performance parameters and limitations.
    - Operating procedures.
- Do I have the required level of proficiency for all phases of the flight taking into consideration the likely conditions to be encountered?
- Am I familiar with the relevant risk reduction strategies and tactics?
- Am I proficient on the airplane type?
- Am I completely familiar with the ATC and any local procedures?
- Am I familiar with the relevant regulatory authority rules and regulations?
- Have I rehearsed the flight in my mind?
- Am I mentally prepared for emergencies?
- Am I dressed (or will be) for the weather outside of the cockpit along the intended route? Can I operate the airplane properly in the clothing being (to be) worn?
- Am I familiar with the available in-flight services?
- Am I familiar with the route maps, airfield details, IFR approach plates, navigation and communication frequencies etc?
- Have I refreshed my knowledge of the likely radio calls?
- Can I readily recall the emergency procedures?
- Have I recently practiced emergency procedures?
- Have I recently practised partial panel flying (IFR pilots)?

Personal Minimums

▸ Is this flight within my personal minimums?
▸ Have I considered all the risk factors?
▸ Have I allowed margins of safety commensurate with my knowledge, experience and proficiency?

Passengers/Co-pilot

▸ If there is a co-pilot, have I predetermined the roles and responsibilities?
▸ Have I forewarned the passengers that a diversion, or a return, is always a possibility?
▸ Do the passengers have appropriate clothing for the flight?

Weather

▸ Have I obtained an appreciation of the overall weather picture?
▸ Have I obtained, assessed and interpreted the forecasts for departure, en-route, destination, alternate and escape routes?
▸ Have I obtained a weather briefing?
▸ Do the forecasts cover the required time periods of the flight?
▸ Have I checked the icing levels and probability of carburetor icing?
▸ Are the conditions suitable for the flight?
▸ Does the weather allow me to return or divert in case of difficulties?
▸ Is the forecast crosswind below the limit specified for the airplane and my current level of proficiency?

Departure

▸ Have I checked the current NOTAM's and other alerts?
▸ Have I obtained and studied the airfield details (including available runway(s), obstructions, recognizable landmarks, local procedures, taxiways and signs)?
▸ Have I checked the required runway length, and compared it to the runway length available, taking into account the surface type and condition, slope, airplane weight, temperature, altitude, wind and other factors that affect the takeoff distance?
▸ Have I checked the runway surface conditions (e.g. wet, dry, snow, ice)?
▸ Have I checked the airplane climb performance and ability to clear any obstacles?
▸ Have I studied the departure approach plates, and are they with me, in case of an IMC return (IFR flight)?

En-Route

- ▶ Have I planned for possible alternative courses of action in case of:
    - ▶ Deteriorating weather?
    - ▶ Equipment failure?
    - ▶ Diversions?
    - ▶ Different departure procedures being required by ATC?
    - ▶ Becoming inadvertently lost (e.g. remote area flying)?
- ▶ Have all NOTAM's been obtained and checked?
- ▶ Have I planned a route and altitudes based on safe practices?
    - ▶ Clear of icing.
    - ▶ At or above lowest safe altitude.
    - ▶ Avoidance of rough and high terrain.
    - ▶ Checked terrain hazards along the route.
    - ▶ Availability of clearly recognizable landmarks (large rivers, lakes, high ground, railways, highways).
- ▶ Have I identified the need for an alternate due to:
    - ▶ Weather?
    - ▶ Runway lighting?
    - ▶ Navigation aids?
- ▶ Have I left sufficient daylight for the flight (day flight)?
- ▶ Have I adequate knowledge of the route to be flown, relevant ATC procedures and restricted/prohibited areas?
- ▶ Have I identified and checked the suitability of any emergency airstrips (runway length, surface, slope)?
- ▶ Have I studied all the relevant information?
- ▶ Have I the correct frequencies for the proposed route?

Fuel

- ▶ Have I allowed sufficient fuel in addition to the normal requirements to cater for:
    - ▶ Possible diversions?
    - ▶ Flight maneuvering?
    - ▶ Variation in fuel consumption?
    - ▶ Alternate?
    - ▶ Holding?
    - ▶ Change in forecast wind?
- ▶ Do I have to limit fuel because of weight, balance or takeoff and landing limitations?
- ▶ Have I allowed sufficient fuel margin consistent with my personal minimums?

Emergency Equipment

- ▶ Do I have relevant equipment in case of an emergency (e.g. life jackets, clothing, portable phone, hand held transceiver, water, food, relevant survival gear)?

Personal Equipment/Data/Information

- ▶ Do I have all the required current maps and documents for the flight?
- ▶ Do I have all the navigation and communication frequencies?
- ▶ Do I have a checklist that is specific for the airplane being flown?
- ▶ Are all the required mandatory documents on board?
- ▶ Does my torch have fresh batteries or do I have spare batteries (for a night flight)?
- ▶ Have I sorted all the documents and organized them for easy access?
- ▶ Is there pen, pencil and paper available as well as stick-on paper in case of instrument malfunction?

Airplane Serviceability

- ▶ Has the airplane been prepared for the likely conditions (e.g. cold weather operation)?
- ▶ Does the airplane have the required serviceable instruments for the flight, in accordance with the relevant regulations?
- ▶ Do I have the equipment needed for the flight?
    - ▶ Emergency equipment.
    - ▶ Tie down equipment.
    - ▶ Spare oil.
- ▶ Is the airplane capable of handling the expected weather?
- ▶ Is the airplane serviceable in every respect?

Weight and Balance

- ▶ Have I completed a weight and balance check, covering departure and arrival?
- ▶ Is the airplane within the correct weight and balance limits?
- ▶ Have I checked the takeoff and landing charts for weight limitations?
- ▶ Can I carry full fuel, passengers and baggage?
- ▶ Are there any weight restrictions because of takeoff and landing limitations or requirements?
- ▶ If the weight and balance is near the limits, have I taken into account the actual weights rather than estimated weights?

Destination and Intermediate Landing Points

> ▸ Have I checked the current NOTAM's and other alerts?
> ▸ Have I obtained permission to use the airfield?
> ▸ Have I obtained and studied the airfield details (including available runway(s), obstructions, recognizable landmarks, local procedures, taxiways and signs)?
> ▸ Do I know the location of the airfield in relation to a town or other prominent landmarks?
> ▸ Have I checked last light?
> ▸ Am I familiar with the local procedures?
> ▸ Have I checked the required runway length, and compared it to the runway length available, taking into account surface type and condition, slope, airplane weight, temperature, altitude, wind and other factors that affect the landing distance?
> ▸ Have I checked the climb gradient and obstacles, in case of a go-around?
> ▸ Have I checked for obstructions, wind direction indication and recognizable landmarks?
> ▸ Is the likely crosswind component within the limit specified for the airplane and my current level of pilot proficiency?
> ▸ Have I checked the runway surface conditions (e.g. wet, dry, snow, ice)?
> ▸ Is the correct grade of fuel available?
> ▸ Am I familiar with the destination and alternate approach plates (IFR flight even if VMC exists)?
> ▸ Are there any conditions that create visual illusions?
> ▸ Is the forecast crosswind below the limit specified for the airplane?
> ▸ Am I proficient at handling the likely conditions (including taxiing)?

Go/No-go Decision

> ▸ Are all the conditions acceptable for the flight, or should it be postponed or cancelled?

Flight Plan

> ▸ Have I completed a Flight Plan?
> ▸ Have I submitted a Flight Plan or Flight Note?
> ▸ Have I completed a navigation and fuel log?

Appendix 3 Typical Single-Engine Airplane Preflight Inspection Checklist

A3

The checklist below is an expanded checklist for single-engine airplanes. It can be used to prepare a customized checklist that includes all items from the POH, and any additional items that may be relevant.

Initial Actions

- Remove tie downs, external control locks, pitot cover, nose plugs, wheel chocks.
- Check that there is no spilt oil or fuel.
- Remove ice, snow or frost from all surfaces or wait until melted.
- Check to see that the airplane is on a level surface and sitting correctly.
- Check that there is no overall damage.
- Check for bird droppings that might be indicative of a nest.

In the Cockpit

Smell	Check that there are no fuel odors (e.g. a leaking compass will leave a kerosene smell).
Control Lock	Remove and stow securely.
Control Column	Move through travel and look/listen to ensure no irregularities (e.g. binding).
Park Brake	On.
Magneto	Switch off and key out.
Avionics Switch	Off.
Master	On.
Retractable Landing Gear	Check that the gear lever is in down position and "down and locked" lights are on.

Turn on pitot heat, anti-collision lights, landing lights and navigation lights. Then leave the cockpit and check the following :

▶	Stall Warning Vane	Check for freedom of movement and buzzer operation.
▶	Pitot Heat	Check that pitot tube is warm (CAUTION: don't burn your hand or flatten the battery).
▶	Lights	Check for correct operation (navigation, landing, taxi, beacon and strobes).

Return to cockpit and continue the check:

▶	Switches	Switch off lights and pitot heat.
▶	Fuel	Check that the fuel is on, and note the fuel tank contents as indicated by the gauges.
▶	Fuel Selector Lever	Check freedom of movement.
▶	Flaps	Lower, and check that flaps are not sluggish or uneven (both manual and electric flaps).
▶	Master	Off.
▶	Trim	Run through full range and set at the neutral position. Check that there is no stickiness, or fouling, of trim controls and cables.
▶	Windshield	Check condition and cleanliness.
▶	Cockpit	Check cleanliness.
▶	Fire Extinguisher	Check in position, secure and serviceable.
▶	First Aid Kit	Check available with required contents.
▶	Seat Rails	Inspect for condition and locking mechanism function.
▶	Placards	Check secure and legible.
▶	Seat Belts	Check security, and that there is no fraying or deterioration. Check that the seat stops are in place, and seat-retaining pins are locked.
▶	Doors	Check for correct fitting, closure, and security.
▶	Oxygen	Check system.

External

Port Fuselage

▶ Covering	Check general condition, and that there are no loose/missing rivets, cracks, tears, wrinkles, unacceptable corrosion or dents.
▶ Radio Aerial	Check secure.
▶ Static Vent	Check clear and unblocked (do not blow into vent).
▶ Drain Holes	Check free of obstructions (know where they are).

Baggage Compartment(s)

▶ Load	Check that the load is secure under a net or strapping and within weight and balance limits.
▶ Baggage Compartment	Check that the door hinges are secure and door(s) closed (but unlocked in case of an emergency).

Empennage

▶ Control Surface Lock	Remove (if not already done).
▶ Fixed/Moveable Surfaces	Check that there are no cracks, loose/missing rivets, corrosion, wrinkles or unacceptable dents. Check rudder and elevator hinges, hinge pins and linkages for condition and security. Check for full freedom of movement, and that there is no excessive play. Check underneath to make sure it is not damaged. Check counterweights are secure.
▶ Navigation Light	Check condition and security.
▶ Trim Tabs	Check in neutral position, secure and free of cracks.
▶ Tail Wheel	Check springs, steering arm and chains.
▶ Lights	Check navigation and anti-collision lights for condition and security.

Starboard Fuselage

▶ As for port fuselage	

Starboard Wing

▸ Control Surface Lock	Remove (if not already done).
▸ Tie Down	Remove (if not already done).
▸ Flaps	Check upper and lower surface for condition, that the linkages are secure and that they are adequately greased. Check for correct level of free play and that there is no excessive wear on the guides (no burring). Check that the rollers (where applicable) are within the guides.
▸ Ailerons	Check upper and lower surface for condition and that linkages, hinge pins and balance weights are secure. Check hinges are not cracked. Check for full and free movement, and confirm that there are no linkage problems. Check that there is no excessive play in control surface hinges.
▸ Bonding Strip	Check secure and not broken.
▸ Wing Tip	Check security, and that there are no cracks, impact damage or damage to the strobe/navigation lights.
▸ Leading Edge	Check cleanliness, that there are no unacceptable dents, and no evidence of over stressing.
▸ Wing Struts	Check secure, and that there is no abnormal rippling around the strut junctions.
▸ Wing Roots	Check that there is no abnormal rippling.
▸ Fuel Tank Vents	Check that the vents are unobstructed (e.g. free of insects/mud) and undamaged.
▸ Wing Surface	Check condition and that there are no missing rivets or unacceptable corrosion, wrinkles or dents. Check freedom from contamination such as frost, mud or snow.

Starboard Landing Gear

▶ Tires	Check that the tires are properly inflated (check the POH for the required pressure) and that there are no cuts, bruises or evidence of excessive wear. Roll airplane back to inspect the entire surface (a smooth tire is unserviceable when the cords are showing). Check the sidewalls of retractable gear tires.
▶ Wheels	Check that there are no cracks.
▶ Gear Well	Check security (gear door and linkages) and that there are no obstructions.
▶ Oleos/Shock Struts	Check cleanliness and proper inflation.
▶ Limit/Position Switches	Check security, cleanliness and condition.
▶ Gear Struts	Check for proper inflation, cleanliness and that there are no leaks or cracks.
▶ Hydraulic Line	Check that there is no damage or leaks.
▶ Wheel Fairing	Check general condition, security, and that they are free of any mud accumulation.
▶ Disc Brakes	Check shiny and not rusted or pitted. Check brake pads for serviceability.
▶ Calipers	Ensure retaining bolts secure.
▶ Chocks/Safety Lock	Remove (if not already done).

Starboard Fuel Tank Drain

▶ Sample	Rock wings before taking a sample to dislodge any trapped water. Take two samples (to ensure all water is drained) and check for contamination, smell and color. Make a visual assessment of the fuel: ▶ Correct color. ▶ Fuel should be clear and bright. ▶ Undissolved water will appear as droplets on the sides, or as bulk water at the bottom of the sample. Suspended water droplets in the fuel will have a cloudy appearance. If water is detected, keep on draining until you get pure fuel. ▶ Solid matter will usually consist of rust, sand, dust or scale. Smell the fuel. It should smell like gasoline and not kerosene – kerosene leaves an oily residue if rubbed between the fingers. Discard sample fuel (do not return to tanks).
▶ Drain Valve	Check completely shut and not leaking.

Starboard Fuel Tank

▶ Fuel Quantity	Physically check fuel quantity with the airplane on level ground (using a properly calibrated dip stick or filled to the tabs or full).
▶ Filler Cap	Check seal(s) are not dry and the gasket is free of cracks. Re-secure the cap. Check that it is the correct type (e.g. vacuum vented). If the tank hisses when you release the cap, there may be a clogged vent. If the fuel cap has a vent, make sure it is not covered with dirt, wax or polish.
▶ Vent	Check free of obstructions.

Cowling and Engine

▶ Inside Engine Compartment	Smell to check that there are no fuel leaks.
▶ Under Cowling	Check that there is no evidence of a bird nest, rags left inadvertently, cracked exhaust system and heater muffs, frayed wires, linkages not properly secure, wear of oil or fuel lines due to metal-to-metal contact or fuel and oil leaks (smell and look). Check alternator drive belt condition, and that the tension is correct.
▶ Oil	Wipe the dipstick, and then check the oil quantity relative to max/min requirements. Check for correct condition (no metal, water contamination and that there is no evidence of poor condition oil rings - oil will look like black lacquer). Add oil as required. Do not over fill. Ensure the dipstick is properly seated and oriented. Re-fit the oil filler cap.
▶ Oil Cooler	Check secure and unobstructed.
▶ Induction Air Filter	Check secure and unobstructed.
▶ Cowling Exterior	Check for loose or missing fasteners.
▶ Crankcase Breather Tube	Check that there is no blockage.
▶ Taxi/Landing Lights	Check condition, cleanliness and security.
▶ Cowl Flaps	Check condition, and security.
▶ Underneath	Check to see that there are no indications of a fuel leak (fuel stains), an exhaust system leak (brownish trails), an oil leak (excessive oil stains), a rich mixture (dry black soot) or leakage past the cylinders (oily black soot).
▶ Exhaust Pipe	Run a finger on the inside of the exhaust pipe, and check that there is no evidence of a rich mixture (dry black soot) or leakage past the cylinders (oily black soot). Check security. CAUTION: Carry out only if cold.

Fuel Strainer/Fuel Drain

▸ Sample	Take sample and check for contamination/water. Note that on some airplanes two samples should be taken – one for each position of the fuel selector lever (check the POH).
▸ Valve	Check completely shut, and not leaking.

Nose Landing Gear

▸ Tire	Check that the tire is properly inflated and that there are no cuts, bruises or evidence of excessive wear. Roll airplane back to inspect the entire surface (a smooth tire is unserviceable when the cords are showing). Check the sidewalls of retractable gear tires.
▸ Hydraulic Lines	Check that there is no damage, and no evidence of leaks.
▸ Wheel Fairing	Check condition and security. Clear any mud accumulation.
▸ Wheel Well	Check cleanliness.
▸ Disc Brakes	Check they are shiny, and not rusted or pitted.
▸ Oleo	Check for proper inflation, cleanliness.
▸ Limit/Position Switches	Check cleanliness, condition and security.
▸ Shimmy Dampers	Check security and that there is no damage.
▸ Chocks	Remove (if not already done).

Propeller

▸ Propeller Blades	Check that the leading edge, trailing edge, front and back surfaces are undamaged, and that there are no unacceptable nicks, cracks, scratches or corrosion. Check that the tip is not bent and that the blades are clean.
▸ Hub and Spinner	Check security, condition and that there is no evidence of oil leakage.
▸ Ground Area	Check free from loose stones.

Note: It is not recommended that the propeller be rotated backwards to check engine compression, as this may damage the vacuum pump.

Port Landing Gear

▸ As for starboard landing gear	

Port Fuel Tank

▸ As for starboard fuel tank

Port Fuel Tank Drain

▸ As for starboard fuel tank drain

Other Fuel Drains

▸ Some airplanes have other drain points   Check the POH and take samples as per the POH.

Port Wing

▸ As for starboard wing
▸ Pitot Tube   Remove pitot tube cover (if not already done). Check that it is clear of obstructions (do not blow into the tube). Check that the drain hole is not blocked.

Cockpit Door(s)

▸ Check that the handles, latches and hinges are secure and there are no loose screws. Also check that the doors close and latch properly. Check condition of the hinges.

Windshield

▸ Check that the windshield is clean (needed for collision avoidance).
▸ Check that there are no signs of oil spray (oil spray could be from a leaking constant speed propeller).

Some Final Questions to Ask

▸ Have the tie downs, pitot tube cover and control locks been removed and stowed?
▸ Have I personally checked the amount of fuel in the tanks?
▸ Have the tanks and filter bowls been checked for water?
▸ Is the oil level correct?
▸ Are the oil cap and dipstick secured?
▸ Is the load properly secured?
▸ Are the fuel tank caps properly sealed and secured?
▸ Are the baggage doors properly secured (but not locked – in case of emergency)?
▸ Is the ELT functioning?
▸ Have I omitted to check anything?

Abbreviations

AB

AGL	Above Ground Level
AI	Attitude Indicator
AIRMET	Airman's Meteorological Advisory
ATC	Air Traffic Control
ATIS	Automatic Terminal Information Service
CAT	Carburetor Air Temperature
CH	Carburetor Heat
CHT	Cylinder Head Temperature
COM	Communications
CRM	Cockpit Resource Management
EGT	Exhaust Gas Temperature
FAA	(United States) Federal Aviation Administration
GA	General Aviation
GPS	Global Positioning System
HI	Heading Indicator
IFR	Instrument Flight Rules
IMC	Instrument Meteorological Conditions
MP	Manifold Pressure
NAV	Navigation
NOTAM	Notice to Airmen
PCATD	Personal Computer-based Aviation Training Device
PIREP	Pilot Report
POH	Pilot Operating Handbook/Aircraft Flight Manual
SIGMET	Significant Weather Report
TAF	Aerodrome Forecast
TC	Turn Coordinator
VFR	Visual Flight Rules
VMC	Visual Meteorological Conditions
VOR	VHF Omni-directional Range

References

Aircraft Owners and Pilots Association, AOPA Pilot:
Ells, Stephen. *Exhausted and Often Forgotten*, March 2001.
Horne, Thomas A. *Storm Warnings*, April 1996.
Schiff, Barry. *Proficient Pilot – Aggressive Safety*, November 2000.

Aircraft Owners and Pilots Association, Air Safety Foundation Safety Advisor, *Propeller Safety*.

Aircraft Owners and Pilots Association, Air Safety Foundation:
Weather Strategies.
Weather Tactics.
*General Aviation Weather Accidents – An Analysis and Preventative Strategie*s.

Australian Aviation Underwriting Pool. *AAUP Insight*
(Volume 7, Number 26), Safety First – Checklists.

Australian Bureau of Meteorology, Manual of Meteorology. (Copyright Commonwealth of Australia. Excerpts reproduced by permission.):
Part I   General Meteorology.
Part II   Aviation Meteorology.

Australian Defence Forces. *Icing and All That*, Directorate of Flight Safety – ADF Flying Feedback, June 1995.

AVweb (The Internet's aviation magazine and news service at http:/www.avweb.com):
Deakin, John. *Throw Away Those Stupid Checklists.*
Printup, Mark. *The Effect of Fatigue on Performance and Safety.*
Puddy, R. Scott. *The Go/No-go Decision.*
Schwaner, John. *How to Monitor Your Engine's Condition.*

Bramson, Alan. *Be a Better Pilot*, Martin Dunnitz Limited, 1980.

Bureau of Air Safety Investigation, Australian Department of Transport and Regional Development, *Asia Pacific Air Safety.*

Bureau of Air Safety Investigation, Australian Transport Safety Bureau, *Human Factors in Fatal Aircraft Accidents*, April 1996.

Bureau of Air Safety Investigation, Australian Department of Transport and Communications, *Limitations of See and Avoid Principle*, April 1992.

Civil Aviation Authority of New Zealand, Good Aviation Practice (GAP):
    Mountain Flying.
    Takeoff and Landing Performance.
    Winter Flying.

Civil Aviation Safety Authority, Australia, *Flight Safety Australia.*

Civil Aviation Safety Authority, Australia, *See and Avoid.*

Department of Civil Aviation (Australia), *Aviation Safety Digest.*

Federal Aviation Administration. Aviation Circulars:
    Carbon Monoxide Contamination in Aircraft Detection and Prevention, AC 20-32.
    Prevention of Retractable Gear Failures, AC 20-34.
    Pilot's Spatial Disorientation, AC 60-4.
    Aeronautical Decision Making, AC 60-22.
    Role of Flight Preparation, AC 61-84.
    Effect of Icing on Aircraft Control and Airplane De-Ice and Anti-Ice Systems, AC 91-51.
    Reciprocating Engine Power Loss Accident Prevention and Trend Monitoring, AC 20-105.
    Aircraft Inspection for General Aviation Aircraft Owners, AC 20-106.
    Pilot Precautions and Procedures to be Taken in Preventing Aircraft Reciprocating Engine Induction System and Fuel System Icing Problems, AC 20-113.

Federal Aviation Administration, Spring 2000, *Dehydration & the Pilot*, Air Surgeons Medical Bulletin.

Federal Aviation Administration, General Aviation Accident Prevention Program:
    Time in Your Tanks, FAA-P-8740-3.
    How to Obtain a Good Weather Briefing, FAA-P-8740-30.
    Thunderstorms – Don't Flirt … Skirt'em, FAA-P-8740-12.
    Engine Operation for Pilots, FAA-P-8740-13.
    Maintenance Aspects of Owning Your Own Aircraft, FAA-P-8740-15.
    Tips on Winter Flying, FAA-P-8740-24.
    Meet Your Aircraft, FAA-P-8740-29.
    Meet Your Aircraft Quiz, FAA-P-8740-29A.
    All About Fuel, FAA-P-8740-35A.
    Proficiency and the Private Pilot, FAA-P-8740-36.
    Human Behaviors, Faa-P-8740-38.
    Winter Survival – The Skill You Never Want to Use, FAA-P-8740-53.

Federal Aviation Administration, *General Aviation Preflight Planning to Reduce Accidents*, FAA SP-94/1-LR.

Federal Aviation Administration, *Over the Counter Medication and Flying*, Medical Facts for Pilots, AM-400-92/1.

Federal Aviation Administration, *Runway Incursions*, FAA/ASY-300.

Fowler, R. *Making Perfect Landings*, Iowa State University Press, 1984.

Fried, Howard. *Beyond the Checkride - What Your Flying Instructor Never Taught You*, McGraw Hill, 1997.

Goss, Richard L. and Editorial Staff. *Studies Suggest Methods for Optimizing Checklist Design and Crew Performance*, Flight Safety Digest, May 1995.

Krause, Shari Stamford. *Collision Avoidance Must Go Beyond "See and Avoid" to "Search and Detect"*, Flight Safety Digest, May 1997.

Transport Canada, *Aviation Safety Letter*:
        4/82 *Fatigue – The Insidious Killer.*
        6/83 *The Major Weather Problems.*

Turner, Thomas P. *Gear-Up Landings*, AOPA (Australia) Magazine, September 2001 (Article originally appeared in the Australian Bonanza Society Newsletter).

UK Civil Aviation Authority. Aeronautical Information Circulars:
        Takeoff, Climb and Landing Performance of Light Aircraft, AIC 12/1996.
        Modern Medical Practices and Flight Safety, AIC 20/1997.
        Medication, Alcohol and Flying, AIC 114/1996.
        The Effect of Thunderstorms and Associated Turbulence on Aircraft Operations, AIC 124/1996.
        Flight Over and in the Vicinity of High Ground, AIC 144/1997.
        Induction System Icing on Piston Engines as Fitted to Aeroplanes, Helicopters and Airships, AIC 145/1997.

UK Civil Aviation Authority, General Aviation Safety Sense Leaflets:
        No 3   *Winter Flying.*
        No 7   *Aeroplane Performance.*
        No 9   *Weight and Balance.*
        No 12   *Strip Sense.*
        No 13   *Collision Avoidance.*
        No 14   *Piston Engine Icing.*
        No 21A   *Ditching.*
        No 23   *Pilots: It's Your Decision.*

Selected Bibliography

B

Aircraft Owners and Pilots Association, *AOPA Pilot*.

Aircraft Owners and Pilots Association, Air Safety Foundation, *Safety Advisor* series.

Buck, Bob. *Weather Flying*, McGraw Hill, 1998

Controlling Pilot Error Series, McGraw Hill, 2000.
> *Automation*, Vladimir Risukhin.
> *Checklists and Compliance*, Thomas P. Turner.
> *Communication*, Paul E. Illman.
> *Controlled Flight into Terrain CFIT/CFTT*, Daryl R. Smith.
> *Culture, Environment and CRM*, Tony Kern.
> *Fatigue*, James C. Miller.
> *Maintenance and Mechanics*, Larry Reithmaier.
> *Situational Awareness*, Paul A. Craig.
> *Training and Instruction*, David E. Frazier.
> *Weather*, Terry T. Lankford

Campbell, R D. *Flight Safety in General Aviation*, Collins Books, 1987.

Helmreich, Robert L and Merritt, Ashleigh C. *Error and Error Management*, University of Texas Aerospace Research Project, Technical Report 98-03, May 19, 1998.

Hurst, R L. *Pilot Error – The Human Factor*, Granada, 1982.

Internet based articles.
> www.pilotfriend.com. *Ditching Aircraft.*
> www.equipped.org. *Ditching.*

Janssens, Leo. *Landing an Iced-Up Airframe*, Avweb 1996 (www.avweb.com).

Jensen, Richard S. *Pilot Judgement and Crew Resource Management*, Avebury Aviation, Ashgate Publishing Limited, 1995.

Krause, Shari Stamford. *Safety – Accident Investigation, Analysis and Application*, McGraw Hill, 1996.

Schiff, Barry. *The Proficient Pilot*: Volume 1, Aviation Supplies and Academics, 1997.

Schiff, Barry. *The Proficient Pilot*: Volume 2, Aviation Supplies and Academics, 2001.

TAB Practical Flying Series, McGraw Hill.
> *Avoiding Mid-Air Collisions*, Shari Stamford Krause, 1995.
> *Cockpit Resource Management: The Private Pilot's guide*, Thomas P. Turner, 1998.
> *Cockpit Weather Decisions*, Terry T. Lankford, 1998
> *Flying in Adverse Conditions*, R. Randall Padfield, 1994.
> *Night Flying*, Richard F. Haines and Courtney L. Flatau, 1992.
> *The Aviators Guide to Flight Planning*, Donald J. Clausing, 1989.

Welch, Anne. *Pilot's Weather – A Flying Manual*, John Murray (Publishers) Ltd, 1973.

Index

J

K

L

M

N

P

R

S

T

U

V

W